Molecular Biology Intelligence Unit

Networking of Chaperones by Co-Chaperones

Gregory L. Blatch, Ph.D.
Department of Biochemistry, Microbiology and Biotechnology
Rhodes University
Grahamstown, South Africa

Landes Bioscience / Eurekah.com
Austin, Texas
U.S.A.

Springer Science+Business Media
New York, New York
U.S.A.

NETWORKING OF CHAPERONES BY CO-CHAPERONES

Molecular Biology Intelligence Unit

Landes Bioscience / Eurekah.com
Springer Science+Business Media, LLC

ISBN: 978-0-387-49308-4 Printed on acid-free paper.

Springer Science+Business Media, LLC, 233 Spring Street, New York, New York 10013, U.S.A.
http://www.springer.com

Please address all inquiries to the Publishers:
Landes Bioscience / Eurekah.com, 1002 West Avenue, 2nd Floor, Austin, Texas 78701, U.S.A.
Phone: 512/ 637 6050; FAX: 512/ 637 6079
http://www.eurekah.com
http://www.landesbioscience.com

Printed in the United States of America.

9 8 7 6 5 4 3 2 1

Library of Congress Cataloging-in-Publication Data

A C.I.P. Catalogue record for this book is available from the Library of Congress.

About the Editor...

GREGORY L. BLATCH is Professor of Biochemistry at Rhodes University, Grahamstown, South Africa. His research interests fall within the broad field of cell stress and chaperones, with a focus on the role of co-chaperones in the regulation and networking of the major molecular chaperones, Hsp70 and Hsp90. He was recently awarded a Wellcome Trust International Senior Research Fellowship for his research on the biomedical aspects of chaperones and co-chaperones. He received his Ph.D. from the University of Cape Town (UCT), South Africa, and did his Postdoctoral at Harvard University Medical School, U.S.A.

CONTENTS

EDITOR

Gregory L. Blatch
Department of Biochemistry, Microbiology and Biotechnology
Rhodes University
Grahamstown, South Africa
Email: g.blatch@ru.ac.za
Chapter 3

CONTRIBUTORS

Karsten Böhse
Institut für Zellbiologie und Bonner Forum Biomedizin
Rheinische Friedrich-Wilhelms Universität Bonn
Bonn, Germany
Chapter 10

Andreas Bracher
Department of Cellular Biochemistry
Max Planck Institute of Biochemistry
Martinsried, Germany
Chapter 1

Graeme Bradley
Department of Biochemistry, Microbiology and Biotechnology
Rhodes University
Grahamstown, South Africa
Email: g.bradley@ru.ac.za
Chapter 3

Jeffrey L. Brodsky
Department of Biological Sciences
University of Pittsburgh
Pittsburgh, Pennsylvania, U.S.A.
Email: jbrodsky@pitt.edu
Chapter 1

Dejan Bursać
Department of Biochemistry and Molecular Biology
Bio21 Institute of Molecular Science and Biotechnology
University of Melbourne
Parkville, Victoria, Australia
Chapter 9

Avrom J. Caplan
Department of Pharmacology and Biological Chemistry
Mount Sinai School of Medicine
New York, New York, U.S.A.
Email: avrom.caplan@mssm.edu
Chapter 5

J. Paul Chapple
Molecular Endocrinology Centre
William Harvey Research Institute
Barts and the London School of Medicine and Dentistry
Queen Mary University of London
London, U.K.
Email: j.p.chapple@qmul.ac.uk
Chapter 11

Michael Cheetham
Division of Molecular and Cellular Neuroscience
Institute of Ophthalmology
London, U.K.
Email: michael.cheetham@ucl.ac.uk
Chapter 11

Marc B. Cox
Department of Biochemistry and Molecular Biology
Mayo Clinic Arizona
Scottsdale, Arizona, U.S.A.
Chapter 2

Peter Csermely
Department of Medical Chemistry
Semmelweis University
Budapest, Hungary
Chapter 3

Douglas M. Cyr
Department of Cell
and Developmental Biology
School of Medicine
University of North Carolina
at Chapel Hill
Chapel Hill, North Carolina, U.S.A.
Email: dmcyr@med.unc.edu
Chapter 4

Sheril Daniel
Department of Biochemistry
Microbiology and Biotechnology
Rhodes University
Grahamstown, South Africa
Chapter 3

Johanna Dudek
Medizinische Biochemie
Universität des Saarlandes
Homburg, Germany
Chapter 8

Henry F. Epstein
Department of Neuroscience
and Cell Biology
University of Texas Medical Branch
at Galveston
Galveston, Texas, U.S.A.
Email: hepstein@utmb.edu
Chapter 6

Markus Genau
Institut für Zellbiologie
und Bonner Forum Biomedizin
Rheinische Friedrich-Wilhelms
Universität Bonn
Bonn, Germany
Chapter 10

Markus Greiner
Medizinische Biochemie
Universität des Saarlandes
Homburg, Germany
Chapter 8

Jörg Höhfeld
Institut für Zellbiologie
und Bonner Forum Biomedizin
Rheinische Friedrich-Wilhelms
Universität Bonn
Bonn, Germany
Email: hoehfeld@uni-bonn.de
Chapter 10

Martin Jung
Medizinische Biochemie
Universität des Saarlandes
Homburg, Germany
Chapter 8

Trevor Lithgow
Department of Biochemistry
and Molecular Biology
Bio21 Institute of Molecular Science
and Biotechnology
University of Melbourne
Parkville, Victoria, Australia
Email: t.lithgow@unimelb.edu.au
Chapter 9

Han Liu
School of Biosciences
University of Birmingham
Birmingham, U.K.
Chapter 7

Peter A. Lund
School of Biosciences
University of Birmingham
Birmingham, U.K.
Email: p.a.lund@bham.ac.uk
Chapter 7

Odutayo O. Odunuga
Department of Neuroscience
and Cell Biology
University of Texas Medical Branch
at Galveston
Galveston, Texas, U.S.A.
Chapter 6

Meredith F.N. Rosser
Department of Cell
and Developmental Biology
School of Medicine
University of North Carolina
at Chapel Hill
Chapel Hill, North Carolina, U.S.A.
Chapter 4

David F. Smith
Department of Biochemistry
and Molecular Biology
Mayo Clinic Arizona
Scottsdale, Arizona, U.S.A.
Email: smith.david26@mayo.edu
Chapter 2

Csaba Söti
Department of Biochemistry
Microbiology and Biotechnology
Rhodes University
Grahamstown, South Africa
Chapter 3

Jacqueline van der Spuy
Division of Molecular
and Cellular Neuroscience
Institute of Ophthalmology
London, U.K.
Chapter 11

Andreas Weitzmann
Medizinische Biochemie
Universität des Saarlandes
Homburg, Germany
Chapter 8

Britta Westhoff
Institut für Zellbiologie und Bonner
Forum Biomedizin
Rheinische Friedrich-Wilhelms
Universität Bonn
Bonn, Germany
Chapter 10

Richard Zimmermann
Medizinische Biochemie
Universität des Saarlandes
Homburg, Germany
Email: bcrzim@med-rz.uni-saarland.de
Chapter 8

PREFACE

There are a number of books dedicated to the cellular and molecular biology of chaperones and their important role in facilitating protein folding; however, this is the first book dedicated to the co-chaperones that regulate them. This book is perhaps long overdue, as the concept of co-chaperones has been in place for more than a decade. The chapters reflect many of the emerging themes in the field of co-chaperone-chaperone biology, with a particular emphasis on the co-chaperones of the major molecular chaperones, Hsp70 and Hsp90.

What constitutes a co-chaperone? In formal terms, a co-chaperone may be defined as any non-substrate protein that interacts specifically with a molecular chaperone and is important for efficient chaperone function. Many co-chaperones are dedicated to a specific chaperone and play a regulatory role (e.g., Hsp40 regulates the nucleotide-bound state of Hsp70). This regulatory role is highly substrate-sensitive, with some co-chaperones having the ability to directly interact with the substrate protein and target it to the chaperone. Indeed, some co-chaperones have the capacity to carry out some of the functions of a chaperone, such as the prevention of protein aggregation (e.g., some Hsp40s, UNC-45 and Cdc37). However, co-chaperones do not always have the ability to interact with substrate or to act as true chaperones in their own right. Nevertheless, whether they directly bind the substrate or indirectly "sense" its presence, in many cases co-chaperones provide specificity to their somewhat promiscuous chaperone partner.

The structure of co-chaperones suggests that they have evolved through domain recruitment, manifesting as highly sophisticated protein scaffolds for the efficient spatial orientation of protein-protein interaction domains (e.g., J domain) and motifs (e.g., tetratricopeptide repeat [TPR] motif). A number of the chapters document the rapidly emerging structural data on domains and motifs, giving us insight into the elegant manner in which these structural features are the functional engines driving the optimal docking and regulation of chaperones by co-chaperones. Interestingly, evidence has also emerged for "fractured" co-chaperones (e.g., Zim17 in yeast), which represent the evolution of physically uncoupled, yet functionally linked, partner domains, allowing for the flexibility of multiple roles.

Contrary to the perception that co-chaperones are merely auxiliary components of the cell's molecular chaperone machinery, a number of chapters suggest that co-chaperones are core components of, and can sometimes transcend, the chaperone machinery (e.g., the role of GrpE as a thermosensor; and Hop may not be dedicated to Hsp70 and Hsp90). Furthermore, co-chaperones not only play an important role in the regulation of Hsp70 and Hsp90 protein folding pathways, but also integrate these folding pathways with protein degradation pathways so as to maintain cellular homeostasis. Therefore, co-chaperones can be broadly viewed as quality control factors enabling the major molecular chaperones to integrate diverse cellular signals and make the correct decision on whether to hold, fold, or degrade; the global safety of the cell being paramount. Finally, the dogma that chaperones interact only with misfolded or denatured substrate proteins is being challenged by mounting evidence to indicate that co-chaperones are able to target chaperones to act with near native proteins to facilitate conformational change (e.g., targeting of clathrin to Hsp70 by auxilin). The name co-chaperone is perhaps limiting, and as more details on the global cellular roles of co-chaperones are revealed, we will no doubt have to re-evaluate the co-chaperone paradigm.

Gregory L. Blatch, Ph.D.

Acknowledgments

I have been very privileged to have had the opportunity to edit the first book dedicated to co-chaperones. Privileged, firstly because it has given me many new and exciting insights into this fascinating field of research, and secondly because it has allowed me to enter into a thoroughly enriching process of interacting with a highly professional network of biologists. Like any typical network, there were many weak links (the email conversations) and a few strong links (the book chapters) in the network of interactions between editor and authors! And so it was that this book on the "Networking of Chaperones by Co-chaperones" was born. I hope that each of the contributors to this book enjoyed the process as much as I have; thank you for your immense creative input. I am also very grateful to the Rhodes University Chaperone Research Group for so eagerly assisting me at the whole book proofing stage: Dr. Aileen Boshoff, Melissa Botha, Sheril Daniel, Dr. Linda Stephens, Dr. Victoria Longshaw, Michael Ludewig, Dr. Eva-Rachele Pesce, Mokgadi Setati and Addmore Shonhai. I went to many people for advice; thanks to all of you for your valuable time, but especially Dr. Graeme Bradley (Rhodes University), Dr. Peter Lund (Birmingham University, U.K.) and my wife Heather Yule.

CHAPTER 1

Nucleotide Exchange Factors for Hsp70 Molecular Chaperones

Jeffrey L. Brodsky* and Andreas Bracher

Abstract

Hsp70 molecular chaperones hydrolyze and re-bind ATP concomitant with the binding and release of aggregation-prone protein substrates. As a result, Hsp70s can enhance protein folding and degradation, the assembly of multi-protein complexes, and the catalytic activity of select enzymes. The ability of Hsp70 to perform these diverse functions is regulated by two other classes of proteins: Hsp40s (also known as J-domain-containing proteins) and Hsp70-specific nucleotide exchange factors (NEFs). Although a NEF for a prokaryotic Hsp70, DnaK has been known and studied for some time, eukaryotic Hsp70s NEFs were discovered more recently. Like their Hsp70 partners, the eukaryotic NEFs also play diverse roles in cellular processes, and recent structural studies have elucidated their mechanism of action.

Introduction

To cope with environmental stresses, such as heat shock, oxidative injury, or glucose-depletion, the expression of a large number of heat shock proteins (Hsps) is induced in all cell types examined. Early work defined these Hsps (some of which are identical to the glucose-responsive proteins, or Grps) by their apparent molecular masses; thus, Hsps with a mass of ~70 kDa became known as Hsp70s, and ~40 kDa Hsps are Hsp40s.[1] Subsequent studies indicated that many Hsps also function as molecular chaperones, factors that aid in the maturation, processing, or sub-cellular targeting of other proteins.

Perhaps the best-studied group of molecular chaperones is the Hsp70s.[2] Hsp70s are found in every organism (with the exception of some *archaea*[3]) and in eukaryotes reside in or are associated with each sub-cellular compartment. Hsp70s are highly homologous to one another and are comprised of three domains: A ~44 kDa amino-terminal ATPase domain, a central ~18 kDa peptide-binding domain (PBD), and a carboxy-terminal "lid" that closes onto the PBD to capture peptide substrates.[4] In some Hsp70s, a short carboxy-terminal amino acid motif also mediates the interaction between Hsp70s and co-chaperones containing tetratrico peptide repeat (TPR) domains (see Chapters by Cox and Smith, and Daniel et al). By virtue of their preferential binding to hydrophobic peptides, Hsp70s retain these aggregation-prone substrates in solution, which in turn permits Hsp70s to enhance: (1) the folding of nascent or temporarily unfolded proteins; (2) the degradation of mis-folded polypeptides; (3) the assembly of multi-protein complexes; and (4) the catalytic activity of enzyme complexes that might require quaternary assembly. It should come as no surprise, then, that Hsp70 over-expression permits the cell to

*Corresponding Author: Jeffrey L. Brodsky—Department of Biological Sciences, 274A Crawford Hall, University of Pittsburgh, Pittsburgh, Pennsylvania 15260, U.S.A. Email: jbrodsky@pitt.edu

Networking of Chaperones by Co-Chaperones, edited by Gregory L. Blatch.

withstand cellular stresses, and that Hsp70s and constitutively expressed Hsp70 homologues, or Hsp70 "cognates" (also known as Hsc70s) play vital roles in cellular physiology.

Hsp70s bind loosely to their peptide substrates when the ATPase domain is occupied by ATP, and tightly when the enzyme is in the ADP-bound state;[5-8] therefore, ADP-ATP exchange is critical for peptide release, and both ATP hydrolysis and nucleotide exchange are accelerated by Hsp70s co-chaperones. Specifically, Hsp40s—which are defined by the presence of an ~70 amino acid "J" domain—enhance ATP hydrolysis (see Chapter by Rosser and Cyr), whereas ADP release is catalyzed by another group of proteins, known as nucleotide exchange factors (NEFs). In fact, these factors do not "exchange" one nucleotide for another, but because ATP is present at much higher concentrations than ADP in the cell, ATP binding most commonly follows ADP release.

The physiological consequences of eukaryotic Hsp70-Hsp40 interaction are well-characterized.[9-11] In contrast, the contributions of Hsp70 NEFs in eukaryotic cell homeostasis are only now becoming apparent. Therefore, the purpose of this review is to summarize briefly what is known about the best-characterized Hsp70 NEF, the bacterial GrpE protein, and then to discuss in greater detail the more recent discovery of eukaryotic NEFs in the cytoplasm and in the endoplasmic reticulum (ER). Particular emphasis will be placed on the molecular underpinnings by which these NEFs function, and on important but unanswered questions in this field of research.

GrpE: The Bacterial Nucleotide Exchange Factor for Hsp70

The replication of the λ bacteriophage genome in *E. coli* requires DNA helicase activity at the origin of replication (*ori*). The helicase is initially inhibited by the λP protein, but the protein is displaced by host-encoded Hsp70 and Hsp40 chaperones, which were first named DnaK and DnaJ, respectively, based on the inability of *dnaK* and *dnaJ* mutants to support λ replication.[12] Another mutant that prevented λ replication is encoded by the *grpE* locus.[13] DnaK-DnaJ-dependent liberation of λP from the *ori* and replication of the phage genome can be recapitulated in vitro, and it was discovered that decreased amounts of DnaK are required in these assays if GrpE is also present.[14,15] This phenomenon results from the fact that GrpE strips ADP from DnaK, and the combination of DnaJ and GrpE synergistically enhances DnaK's ATPase activity in single-turnover measurements by 50-fold, or even up to 5000-fold, depending on whether GrpE is saturating.[8,16] The DnaK-DnaJ-GrpE "machine" not only regulates multi-protein complex assembly—as observed during phage λ replication—but assists in the folding of newly synthesized and unfolded polypeptides, and homologues of each of these proteins reside in the mitochondria and help drive the import or "translocation" and maturation of nascent polypeptides in this organelle (see Chapter by Bursać and Lithgow).[17,18]

The Discovery of Hsp70 Nucleotide Exchange Factors in Eukaryotes: Fishing Pays Off

The cytoplasm and ER lumen in eukaryotes contain several Hsp70 and Hsp40 homologues, and it was assumed that GrpE homologues would also reside in these compartments. After many years, the failure to identify them was ascribed either to the fact that GrpE homologues are highly divergent and/or that the Hsp70s in the ER and eukaryotic cytoplasm might have evolved such that GrpE-assisted ADP release is dispensable.[19] Thus, it came as a complete surprise when BAG-1—which was first identified as a cellular partner for Bcl-2, a negative regulator of apoptosis[20]—was found to catalyze ADP release from mammalian Hsp70.[21] The binding between BAG-1 and the ATPase domain of Hsp70 is mediated by a ~50 amino acid "BAG" domain,[22-24] which is present in each of the many isoforms and splice variants of BAG-1 that have been identified. However, it is clear that BAG domain-containing NEFs do not function identically to GrpE, at least in part because their structures are distinct (also see below). For example, GrpE catalyzes the release of both ADP and ATP from DnaK, whereas BAG-1 triggers only ADP release.[25] In addition, GrpE augments DnaK-DnaJ-mediated protein folding and assembly, whereas BAG-1 has been found to exert either positive or negative effects on Hsp70-Hsp40-directed protein folding and chaperone activity. These contradictory results stem primarily from the concentrations of BAG-1 employed and the presence or absence of specific co-chaperones.[26,27] Thus, future work is needed to define how BAG domain-containing proteins impact known chaperone activities and how each of the various isoforms function under normal, cellular conditions and at their native concentrations.

For some time it was thought that yeast lacked a BAG domain-containing protein, but the available structure of an Hsp70 ATPase domain in complex with a BAG domain fragment[28] brought about the discovery of a highly divergent BAG-1 homologue in the yeast database, Snl1.[29] *SNL1* was originally identified as a high-copy suppressor of the toxicity produced by the C-terminal fragment of a nuclear pore protein, and one consequence of this fragment is the generation of nuclear membrane "herniations".[30] Therefore, it was proposed that Snl1 modulates nuclear pore complex (NPC) integrity, and consistent with this hypothesis, Snl1 is an integral membrane protein that resides in the nuclear envelope/ER membrane. Proof that Snl1 is a bona fide BAG homologue derived from the fact that Snl1 associates with Hsp70s from yeast and mammals, and that a purified soluble fragment of Snl1 stimulates Hsp40-enhanced ATP hydrolysis by Hsp70 to the same extent as a mammalian BAG domain-containing protein.[29]

Because the lumen of the ER houses a high concentration of Hsp70 and because of its prominent role in catalyzing the folding of nascent proteins, it was also assumed that a NEF would reside in this compartment. Almost all secreted proteins associate with BiP, the ER lumenal Hsp70, during translocation and folding.[31] During translocation, BiP is anchored to an integral membrane J-domain-containing protein, but if the subsequent folding of the nascent secreted protein is compromised, BiP interacts instead with soluble Hsp40s to facilitate the "retro-translocation" of the aberrant protein from the ER and into the cytoplasm where it is degraded by the proteasome.[32] This process was termed ER associated degradation (ERAD[33]) and is conserved amongst all eukaryotes.

To identify BiP partners that might include NEFs and that might facilitate protein translocation, folding, and/or ERAD, genetic selections were performed in different yeasts. First, the *SLS1* gene was identified in a synthetic lethal screen in *Y. lipolytica* strains that lacked a component of the signal recognition particle, which is essential in this organism for protein translocation.[34] Later studies established that the Sls1 homologue in *S. cerevisiae* interacts preferentially with the ADP-bound form of BiP, that Sls1 enhances the Hsp40-mediated stimulation of BiP's ATPase activity, and that Sls1 accelerates the release of ADP and ATP from BiP.[35] Second, Stirling and colleagues isolated a gene that at high-copy number suppressed a growth defect in *S. cerevisiae* lacking an Hsp70-related protein, known as Lhs1, and that were unable to mount an ER stress response.[36] The gene, *SIL1*, is identical to *SLS1*, and the Sil1 protein was shown to bind selectively to BiP's ATPase domain. Together, these data suggested strongly that Sls1/Sil1 is a BiP NEF. Further support for this hypothesis was provided by the discovery that Sls1/Sil1 is the yeast homologue of BAP, a resident of the mammalian ER that strips nucleotide from BiP and synergistically enhances the J-domain-mediated activation of BiP's ATPase activity.[37]

Surprisingly, Lhs1, mentioned above as an Hsp70-related protein, also appears to function as a NEF. Lhs1 is a member of the Hsp110/Grp170 family of mammalian molecular chaperones that possess N-terminal ATP binding domains with some homology to the Hsp70 ATPase domain; however, the C-terminal halves are comprised of extended, nonconserved polypeptide binding domains.[38] Recent studies from the Stirling laboratory indicate that Lhs1 interacts with BiP in the yeast ER and can strip ADP/ATP from BiP as efficiently as Sls1/Sil1, thus activating BiP's steady-state ATPase activity when combined with a J-domain-containing protein.[39] In turn, BiP activates the ATPase activity of Lhs1, and in both cases the ATP-binding properties of the chaperones are essential for activity. These results indicate that BiP and Lhs1 reciprocally enhance one another's activities, perhaps to coordinate the transfer of polypeptide substrates. Although it is not yet clear whether all members of the Hsp110/Grp170 family are NEFs, another group reported that Hsc70 activates the ATPase activity of a cytosolic, mammalian Hsp110 homologue, Hsp105α, and that Hsp105α inhibits the hydrolysis of ATP-bound Hsc70. These results are consistent with Hsp105α possessing NEF activity.[40]

To identify new cytoplasmic NEFs, we searched the *S. cerevisiae* genome for Sls1 homologues that lacked an ER-targeting sequence and isolated the *FES1* gene.[41] Purified Fes1 catalyzes the release of ADP and ATP from cytoplasmic Hsp70, and the *fes1* thermosensitive growth phenotype is rescued by mutations in a cytoplasmic Hsp40. This genetic finding is consistent with the opposing effects of Hsp40s and NEFs on the identity of the Hsp70-bound nucleotide; i.e., Hsp40s drive Hsp70s into the ADP-bound state, whereas NEFs drive Hsp70s into the ATP-bound state. Interestingly, a mammalian Fes1 homolog—known as HspBP1—was identified previously as a Hsp70 interactor in a yeast two-hybrid screen.[42] Initially, HspBP1 was reported to inhibit nucleotide binding and chaperone activity, but subsequent work by our groups established that HspBP1 also catalyzes nucleotide release from Hsp70.[43,44]

Hsp70 NEFs in Eukaryotes Exhibit Diverse Functions

Hsp70s play a prominent role in many cellular processes, and so it was anticipated that the NEFs would also exhibit diverse functions. Thus far, this prediction has been affirmed, but because this field is in its infancy, relatively little is known, and in some cases—as mentioned above for BAG-1—contradictory results have been obtained. In this section we will highlight key findings, direct the reader to the pertinent literature, and speculate on important directions for future studies.

BAG-1 is a positive or negative regulator of chaperone-mediated protein folding, depending on several variables, and to a large extent these contradictory results derive from the use of in vitro assays in which the experimental conditions may vary from the cellular environment and from in vivo expression systems in which super-stoichiometric amounts of wild type or mutant versions of the protein are produced.[26,27,45] Therefore, and as noted above, future studies must employ conditions that more closely mimic those found in the cell. Nevertheless, what is becoming increasingly clear is that BAG-1 can target proteins for proteasome-mediated degradation (see Chapter by Höhfeld et al). This attribute results from an embedded ubiquitin-like domain in BAG-1,[46] which facilitates proteasome interaction. Because BAG-1 also binds Hsp70, it has been proposed that BAG-1 couples Hsp70 to the proteasome to facilitate chaperone-mediated "decisions" during cytoplasmic protein turn-over. In addition to its role in protein degradation, BAG-1 protects cells against apoptosis, consistent with the association between BAG-1 and Bcl-2. BAG-1 is also involved in androgen receptor and transcriptional activation, and associates with and regulates the Raf-1/ERK kinase. Interestingly, some of these activities are independent of the BAG domain, and thus each BAG-1 homologue probably evolved unique functional motifs to diversify its functions. In addition, these data suggest that BAG domain-containing proteins might prove to be targets for pharmacological interventions to treat human diseases.

The discovery of a yeast BAG-1 homologue, Snl1,[29] provides researchers with a genetic tool to define better how one member of this protein family functions in the cell. As noted above, Snl1 is thought to stabilize the NPC and perhaps modulate its activity,[30] but to date it is not clear how this occurs. Of additional interest is Snl1's localization at the ER membrane, suggesting that the protein might aid Hsp70 and Hsp40 homologues during translocation or ERAD; however, we have found that translocation and ERAD are robust in yeast deleted for *SNL1* either alone or when combined with *fes1* mutants (J. Bennett, J. Young, and G.L. Blatch, unpublished observations).

In contrast, several lines of evidence suggest that the ER lumenal NEF in yeast, Sls1/Sil1, is involved in ERAD and translocation. First, the mRNA encoding Sls1/Sil1 rises when cells are exposed to stresses that activate the unfolded protein response (UPR).[47] Other UPR targeted genes include chaperones and enzymes required for protein folding, post-translational modification, and ERAD, and deletion of *SLS1/SIL1* in one *S. cerevisiae* strain background modestly compromises ERAD efficiency.[47] Second, yeast deleted both for *LHS1* (see above) and for *SLS1/SIL1* exhibit strong translocation defects, although more modest translocation defects are evident in *lhs1*Δ cells.[36] Third, *Y. lipolytica* strains expressing a truncated form of Sls1 that is unable to interact with BiP are translocation-defective.[48] One explanation for each of these findings is that the NEF simply increases the efficiency at which BiP functions during translocation and ERAD, although this has not been demonstrated directly. It will also be vital in the future to determine whether the mammalian homologue, BAP,[37] plays a role in any of these processes.

If BAG-1 and Snl1 are NEFs for cytoplasmic Hsp70s in eukaryotes, why does the cytoplasm harbor the Fes1/HspBP1 proteins? One possibility is that each NEF acts on only a unique Hsp70 or family of Hsp70s. For example, there are seven Hsp70s in the cytoplasm of *S. cerevisiae* that are grouped into distinct classes: One class (the "Ssas") facilitates translocation and ERAD, and others (the "Ssbs" and "Ssz") associate with the ribosome and are involved in translation.[31,49] Although this hypothesis still needs to be examined more thoroughly, we reported that *fes1* mutants display phenotypes consistent with defects in translation initiation and that the Fes1 protein is associated with the ribosome, even though Fes1 is a NEF for an Ssa family member.[41] Yeast deleted for *FES1* also exhibit defects in the folding of newly synthesized firefly luciferase,[44,50] a process that is similarly dependent on the Ssa chaperones.[51,52] Although preliminary, these data suggest that NEFs might be promiscuous when choosing their Hsp70 partners. Otherwise, little else is known about Fes1 homologues except that the levels of HspBP1 are elevated in tumor cells,[53] a result that is consistent with the observation that many tumors contain increased levels of Hsp70.[54] Clearly, much more work is needed on the roles played by Fes1/HspBP1 family members in the cell, an undertaking that will benefit from the construction of new mutants and assays in which their functions can be better defined.

The Mechanism of Action of Hsp70 Nucleotide Exchange Factors: Results from Structural and Biochemical Studies

The first Hsp70 NEF structure determined was the bacterial GrpE in complex with the ATPase domain of its associated Hsp70, DnaK of *E. coli* (Fig. 1A).[55] In the crystal structure and in solution, GrpE forms tight dimers that asymmetrically contact only one ATPase domain.[56,57] GrpE has a bipartite structure composed of an alpha-helical N-terminal part and a small beta-sheet domain at the C-terminus. The alpha-helical fragment forming the dimer interface extends far beyond the measures of the ATPase domain and might contact the substrate-binding region of DnaK. Indeed, whereas full-length GrpE interferes with substrate binding, GrpE missing 33 residues at the N-terminus does not. The interaction with the ATPase domain of DnaK is mediated primarily by the beta-sheet region of one GrpE molecule inserting into the cleft between subdomains IB and IIB of the ATPase domain. The highly conserved ATPase domain of Hsp70/Hsc70/DnaK has a bilobal structure that is conventionally divided further into four subdomains, IA and IB forming lobe I, and IIA and IIB lobe II, respectively.[58] The ATP binding site is located at the bottom of a cleft between subdomains IB and IIB close to the center of the domain. In the structure of the ADP complex of the ATPase domain of mammalian Hsp70, residues from all four subdomains contact the nucleotide. Comparison of the GrpE-DnaK complex with this structure indicated that binding of GrpE induces a 14° rotation of subdomain IIB, resulting in an opening of the nucleotide binding cleft incompatible with nucleotide binding.

The BAG domain of BAG-1 assumes a structure completely unrelated to GrpE, forming a ~60 Å long three-helix bundle, both in solution and in complex with the ATPase domain of Hsc70 (Fig. 1B).[59,60] In the complex, highly conserved polar residues in helices 2 and 3 contact subdomains IB and IIB of the ATPase domain. The majority of interactions are, however, formed with subdomain IIB.[61] The binding of BAG locks the ATPase domain of Hsc70 in a conformation very similar to DnaK in complex with GrpE, with subdomain IIB

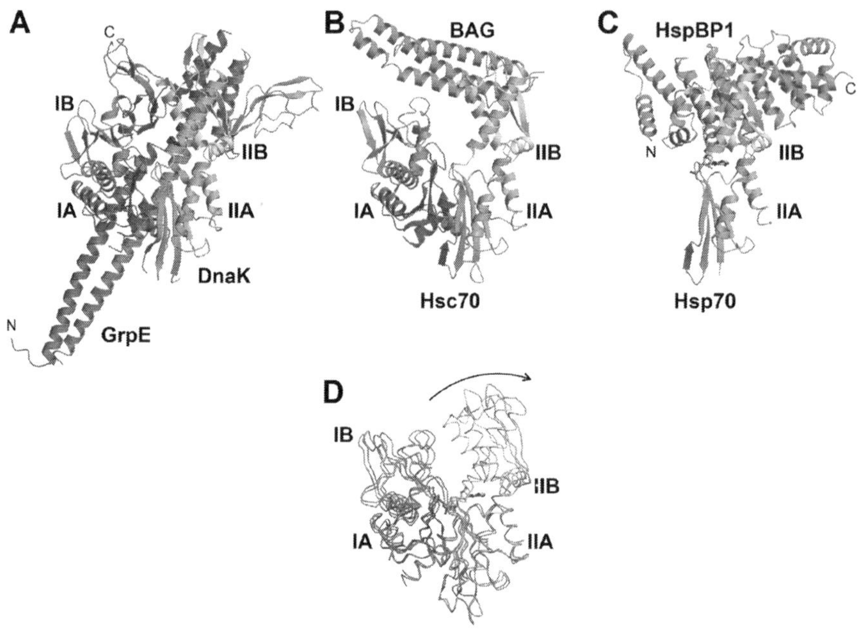

Figure 1. Comparison of the Hsp70 nucleotide exchange factor structures. Panels A-C depict the crystal structures of the complexes of GrpE-DnaK, BAG-Hsc70 and HspBP1-Hsp70, respectively.[44,55,60] The peptide backbones are shown in ribbon representation with the nucleotide exchange factors in green and the ATPase subdomains IA, IB, IIA and IIB in brown, blue, yellow and grey, respectively. For better comparison, the Cα atoms of the ATPase domains were aligned. The HspBP1-Hsp70 structure in panel C contains an additional nucleotide shown in ball-and-stick representation. Augmentation with AMP-PNP was necessary for crystallization of HspBP1 with lobe II of the Hsp70 ATPase domain, but strongly inhibited HspBP1 binding to the full ATPase domain.[44] Panel D illustrates the rotation of subdomain IIB observed between the crystal structures of the ATPase domain in complex with ADP and with BAG.[58,60] The ATPase domains were superimposed, and are shown in the same orientation as in panel B; the peptide backbone of subdomain IIB in the BAG-Hsc70 complex is highlighted in bright green, otherwise the same coloring scheme as in panel B was applied. The figure was created using the programs Molscript and Raster3D.[70,71] A color version of this figure is available online at http://www.eurekah.com.

rotated outward by 14°. These data suggest convergent evolution of the NEFs and are analogous to the structurally divergent nucleotide exchange factors of small G-proteins, all of which employ a common structural switch.[62] Although the ATPase sequences are highly conserved in the Hsp70 family, BAG-1 and GrpE do accelerate nucleotide exchange exclusively on their respective binding partners Hsc70/Hsp70 and DnaK, and it is important to note that the sequences of the ATPase domains of the inducible Hsp70 and the constitutive Hsc70 are virtually identical.[63]

HspBP1, a member of the third class of Hsp70 nucleotide exchange factors, is again structurally unrelated to both GrpE and the BAG domain (Fig. 1C). The core domain, which is sufficient for Hsp70 binding, is composed entirely of alpha-helical repeats containing four regular Armadillo repeats in the central region.[44] Armadillo repeats comprise three helices arranged in an open triangle and are found in many functionally unrelated eukaryotic proteins as a versatile structural building block. In the crystal structure of the complex with lobe II of the Hsp70 ATPase domain, the slightly curved core domain of HspBP1 embraces

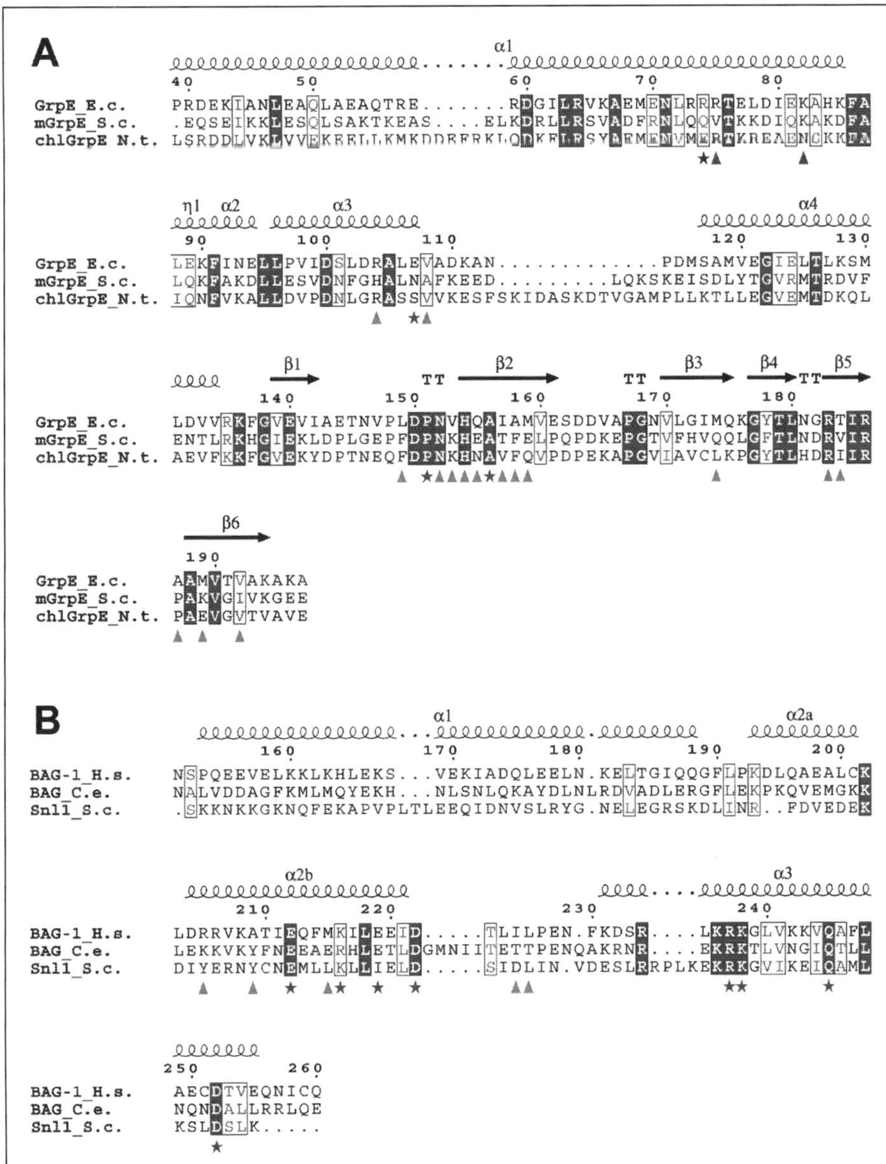

Figure 2A,B. Amino acid sequence alignments of select nucleotide exchange factors. A) Alignment of the structured portion of *E. coli* GrpE with mitochondrial GrpE from *S. cerevisiae* and with chloroplast type-I GrpE from *N. tabacum*. B) Alignment of the BAG domains of human BAG-1, the BAG homolog from *C. elegans*, and Snl1 from *S. cerevisiae*. A color version of this figure is available online at http://www.eurekah.com.

subdomain IIB.[44] By comparison with BAG-1, HspBP1 binds sideways onto subdomain IIB, which would generate a steric conflict of its N-terminus with subdomain IB if the ATPase domain adopts a similar conformation as in the complexes with ADP or BAG-1. Indeed, probing the conformation of the entire ATPase domain in complex with HspBP1 by

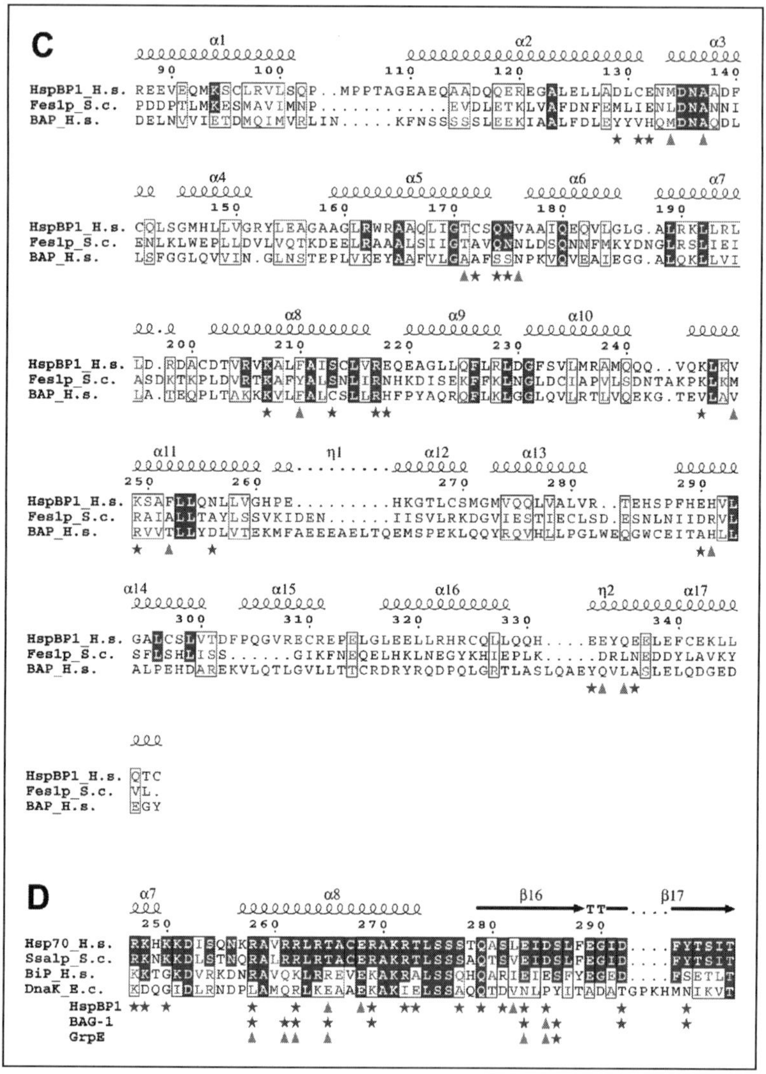

Figure 2C,D. Amino acid sequence alignments of select nucleotide exchange factors. C) Alignment of the core domains of human HspBP1, Fes1 from *S. cerevisiae*, and human BAP. D) Alignment of partial sequences from subdomain IIB of Hsp70 homologs comprising the main binding interface with BAG-1 and HspBP1. GrpE also interacts extensively with subdomain IB. Similar residues are shown in red letters, and identical residues either overall or in a sensible subgroup, e.g., cytosol-ER or eukaryotic-prokaryotic, are shown on a red background. The secondary structure assignment and the numbering correspond to the top sequence in all alignments. Residues involved in polar and van-der-Waals contacts between NEFs and Hsp70 homologs are indicated respectively by red asterisks and green triangles. Contacts in DnaK-GrpE apart from subdomains IB and IIB are shaded in different colors. Accession codes for the sequences are: Q7ABI1, *E. coli* GrpE; P38523, *S. cerevisiae* mGrpE; Q9ZSP4, *N. tabacum* chlGrpE; Q99933, human BAG-1; Q44739, *C. elegans* BAG; P40548, *S. cerevisiae* Snl1; O95351, human HspBP1; NP_009659, *S. cerevisiae* Fes1; AF547994, human BAP; P08107, human Hsp70; P10591, *S. cerevisiae* Ssa1p; P11021, *H. sapiens* BiP; P0A6Z0, *E. coli* DnaK. The alignments were drawn using the program ESPript.[72] A color version of this figure is available online at http://www.eurekah.com.

limited proteolysis and fluorescence spectroscopy suggests a less compact conformation for the ATPase domain as compared to the complex with BAG or in the absence of NEFs. It is thus likely that HspBP1 and its homologs trigger nucleotide exchange by a mechanism distinct from BAG domain proteins and GrpE. The distortion of the Hsp70 ATPase domain might be sufficient to dissociate bound ADP, however rotation of subdomain IIB as observed in the BAG-Hsc70 complex might also occur. The different conformations imposed on the ATPase domain of Hsp70/Hsc70 by HspBP1 and BAG-1 may also differentially affect crosstalk between the substrate binding domain and downstream effectors, like CHIP.[64-66] Because each Hsp70 NEF class is conserved to varying extents (Fig. 2), it is likely that NEF homologs of those for which structures have been determined will function similarly.

Conclusions

As outlined above, initial characterizations and structural studies of eukaryotic Hsp70 NEFs have proceeded rapidly since the relatively recent identification of this family of co-chaperones. What has been more difficult to discern, however, is the spectrum of cellular activities engineered by these proteins. Other Hsp70 co-chaperones appear to augment a sub-set of cellular activities that are normally carried-out by the chaperone (see other chapters in this volume), and we predict that the same rule will apply to Hsp70 NEFs. To some extent this prediction has been borne-out, since unique NEFs facilitate protein folding, translocation, and translation. However, only a relatively small number of Hsp70-catalyzed activities have been examined in these initial studies, and in some cases heterologous reporters (e.g., firefly luciferase) were employed. It is thus imperative that novel cellular assays using endogenous substrates are developed in which the effects of depleting or mutating specific NEFs can be investigated. To this end, the described structural analyses will surely improve our ability to mutate Hsp70-interacting residues on distinct NEFs, and the resulting mutated proteins can then be examined in both genetic (i.e., yeast) and mammalian systems.

Another feature of eukaryotic NEFs that remains mysterious but that will likely become an active area of research is whether these proteins contain built-in stress sensors. Previous work established that the paired, N-terminal helices in *E. coli* GrpE dimers undergo a reversible transition at ~48°C, and that the transition reduces nucleotide exchange activity or association with DnaK. As a result the steady-state population of DnaK becomes predominantly associated with ADP and bound tightly to peptide substrates at elevated temperatures.[67] More recent data indicate that this "thermosensor" is important for the DnaK-DnaJ-GrpE-mediated prevention of protein aggregation and protection of enzyme activity after heat shock.[68] Therefore, GrpE function is one component of the cellular "thermometer" that controls protein folding in the cell. Although the eukaryotic NEFs discussed in this review lack homology to GrpE, it will be interesting to examine whether eukaryotic NEF activities are similarly regulated by temperature or other stresses in vivo.

Finally, given the importance of the Hsp70 chaperone system in human physiology and medicine, we predict that Hsp70 NEFs will emerge as important players in maintaining cellular homeostasis. In turn, we anticipate that defects in the activities of select NEFs will be implicated in disease. Recent data support this supposition: Mice have been found that contain a spontaneous, recessive mutation in the gene encoding a Sil1 (Sls1) homologue.[69] These "woozy" (*wz*) mice accumulate protein inclusions in the ER and nucleus of Purkinje cells and thus become ataxic. Consistent with a role for the murine NEF in protein quality control, the UPR is induced in Purkinje cells from *wz* mutants. The discovery of the *wz* mutation likely represents only the first of many examples in which loss of a NEF homologue in mammals leads to a specific disease or disease-like phenomenon. Therefore, we also predict that the "hunt" will be on for other mutations in mammalian Hsp70 NEFs that impact cellular homeostasis.

References

1. Lee AS. Coordinated regulation of a set of genes by glucose and calcium ionophores in mammalian cells. Trends Biochem Sci 1987; 12:20-23.
2. Mayer MP, Bukau B. Hsp70 chaperones: Cellular functions and molecular mechanism. Cell Mol Life Sci 2005; 62(6):670-684.
3. Macario AJ, de Macario EC. The archaeal molecular chaperone machine: Peculiarities and paradoxes. Genetics 1999; 152(4):1277-1283.
4. Zhu X, Zhao X, Burkholder WF et al. Structural analysis of substrate binding by the molecular chaperone DnaK. Science 1996; 272(5268):1606-1614.
5. Palleros DR, Shi L, Reid Kl et al. hsp70-protein complexes. Complex stability and conformation of bound substrate protein. J Biol Chem 1994; 269(18):13107-13114.
6. Schmid D, Baici A, Gehring H et al. Kinetics of molecular chaperone action. Science 1994; 263(5149):971-973.
7. Prasad K, Heuser J, Eisenberg E et al. Complex formation between clathrin and uncoating ATPase. J Biol Chem 1994; 269(9):6931-6939.
8. McCarty JS, Buchberger A, Reinstein J et al. The role of ATP in the functional cycle of the DnaK chaperone system. J Mol Biol 1995; 249(1):126-137.
9. Kelley WL. The J-domain family and the recruitment of chaperone power. Trends Biochem Sci 1998; 23(6):222-227.
10. Cheetham ME, Caplan AJ. Structure, function and evolution of DnaJ: Conservation and adaptation of chaperone function. Cell Stress Chaperones 1998; 3(1):28-36.
11. Walsh P, Bursac D, Law YC et al. The J-protein family: Modulating protein assembly, disassembly and translocation. EMBO Rep 2004; 5(6):567-571.
12. Yochem J, Uchida H, Sunshine M et al. Genetic analysis of two genes, dnaJ and dnaK, necessary for Escherichia coli and bacteriophage lambda DNA replication. Mol Gen Genet 1978; 164(1):9-14.
13. Saito H, Uchida H. Initiation of the DNA replication of bacteriophage lambda in Escherichia coli K12. J Mol Biol 1977; 113(1):1-25.
14. Alfano C, McMacken R. Ordered assembly of nucleoprotein structures at the bacteriophage lambda replication origin during the initiation of DNA replication. J Biol Chem 1989; 264(18):10699-10708.
15. Zylicz M, Ang D, Liberek K et al. Initiation of lambda DNA replication with purified host- and bacteriophage-encoded proteins: The role of the dnaK, dnaJ and grpE heat shock proteins. EMBO J 1989; 8(5):1601-1608.
16. Liberek K, Marszalek J, Ang D et al. Escherichia coli DnaJ and GrpE heat shock proteins jointly stimulate ATPase activity of DnaK. Proc Natl Acad Sci USA 1991; 88(7):2874-2878.
17. Hohfeld J, Hartl FU. Post-translational protein import and folding. Curr Opin Cell Biol 1994; 6(4):499-509.
18. Bukau B, Horwich AL. The Hsp70 and Hsp60 chaperone machines. Cell 1998; 92(3):351-366.
19. Deloche O, Ang D, Georgopoulos C. The GrpE family of proteins—an overview. In: Gething MJ, ed. Guidebook to Molecular Chaperones and Protein Folding Catalysts. Oxford University Press, 1997:133-137.
20. Takayama S, Sato T, Krajewski S et al. Cloning and functional analysis of BAG-1: A novel Bcl-2-binding protein with anti-cell death activity. Cell 1995; 80(2):279-284.
21. Hohfeld J, Jentsch S. GrpE-like regulation of the hsc70 chaperone by the anti-apoptotic protein BAG-1. EMBO J 1997; 16(20):6209-6216.
22. Takayama S, Bimston DN, Matsuzawa S et al. BAG-1 modulates the chaperone activity of Hsp70/Hsc70. EMBO J 1997; 16(16):4887-4896.
23. Takayama S, Xie Z, Reed JC. An evolutionarily conserved family of Hsp70/Hsc70 molecular chaperone regulators. J Biol Chem 1999; 274(2):781-786.
24. Zeiner M, Gebauer M, Gehring U. Mammalian protein RAP46: An interaction partner and modulator of 70 kDa heat shock proteins. EMBO J 1997; 16(18):5483-5490.
25. Brehmer D, Rudiger S, Gassler CS et al. Tuning of chaperone activity of Hsp70 proteins by modulation of nucleotide exchange. Nat Struct Biol 2001; 8(5):427-432.
26. Takayama S, Reed JC. Molecular chaperone targeting and regulation by BAG family proteins. Nat Cell Biol 2001; 3(10):E237-241.
27. Alberti S, Esser C, Hohfeld J. BAG-1—a nucleotide exchange factor of Hsc70 with multiple cellular functions. Cell Stress Chaperones 2003; 8(3):225-231.
28. Sondermann H, Scheufler C, Schneider C et al. Structure of a Bag/Hsc70 complex: Convergent functional evolution of Hsp70 nucleotide exchange factors. Science 2001; 291(5508):1553-1557.
29. Sondermann H, Ho AK, Listenberger LL et al. Prediction of novel Bag-1 homologs based on structure/function analysis identifies Snl1p as an Hsp70 co-chaperone in Saccharomyces cerevisiae. J Biol Chem 2002; 277(36):33220-33227.

30. Ho AK, Raczniak GA, Ives EB et al. The integral membrane protein snl1p is genetically linked to yeast nuclear pore complex function. Mol Biol Cell 1998; 9(2):355-373.
31. Fewell SW, Travers KJ, Weissman JS et al. The action of molecular chaperones in the early secretory pathway. Annu Rev Genet 2001; 35:149-191.
32. Nishikawa SI, Fewell SW, Kato Y et al. Molecular chaperones in the yeast endoplasmic reticulum maintain the solubility of proteins for retrotranslocation and degradation. J Cell Biol 2001; 153(5):1061-1070.
33. McCracken AA, Brodsky JL. Assembly of ER-associated protein degradation in vitro: Dependence on cytosol, calnexin, and ATP. J Cell Biol 1996; 132(3):291-298.
34. Boisrame A, Beckerich JM, Gaillardin C. Sls1p, an endoplasmic reticulum component, is involved in the protein translocation process in the yeast Yarrowia lipolytica. J Biol Chem 1996; 271(20):11668-11675.
35. Kabani M, Beckerich JM, Gaillardin C. Sls1p stimulates Sec63p-mediated activation of Kar2p in a conformation-dependent manner in the yeast endoplasmic reticulum. Mol Cell Biol 2000; 20(18):6923-6934.
36. Tyson JR, Stirling CJ. LHS1 and SIL1 provide a lumenal function that is essential for protein translocation into the endoplasmic reticulum. EMBO J 2000; 19(23):6440-6452.
37. Chung KT, Shen Y, Hendershot LM. BAP, a mammalian BiP-associated protein, is a nucleotide exchange factor that regulates the ATPase activity of BiP. J Biol Chem 2002; 277(49):47557-47563.
38. Easton DP, Kaneko Y, Subjeck JR. The hsp110 and Grp1 70 stress proteins: Newly recognized relatives of the Hsp70s. Cell Stress Chaperones 2000; 5(4):276-290.
39. Steel GJ, Fullerton DM, Tyson JR et al. Coordinated activation of Hsp70 chaperones. Science 2004; 303(5654):98-101.
40. Yamagishi N, Ishihara K, Hatayama T. Hsp105alpha suppresses Hsc70 chaperone activity by inhibiting Hsc70 ATPase activity. J Biol Chem 2004; 279(40):41727-41733.
41. Kabani M, Beckerich JM, Brodsky JL. Nucleotide exchange factor for the yeast Hsp70 molecular chaperone Ssa1p. Mol Cell Biol 2002; 22(13):4677-4689.
42. Raynes DA, Guerriero Jr V. Inhibition of Hsp70 ATPase activity and protein renaturation by a novel Hsp70-binding protein. J Biol Chem 1998; 273(49):32883-32888.
43. Kabani M, McLellan C, Raynes DA et al. HspBP1, a homologue of the yeast Fes1 and Sls1 proteins, is an Hsc70 nucleotide exchange factor. FEBS Lett 2002; 531(2):339-342.
44. Shomura Y, Dragovic Z, Chang HC et al. Regulation of Hsp70 function by HspBP1: Structural analysis reveals an alternate mechanism for Hsp70 nucleotide exchange. Mol Cell 2005; 17(3):367-379.
45. Gehring U. Biological activities of HAP46/BAG-1. The HAP46/BAG-1 protein: Regulator of HSP70 chaperones, DNA-binding protein and stimulator of transcription. EMBO Rep 2004; 5(2):148-153.
46. Luders J, Demand J, Hohfeld J. The ubiquitin-related BAG-1 provides a link between the molecular chaperones Hsc70/Hsp70 and the proteasome. J Biol Chem 2000; 275(7):4613-4617.
47. Travers KJ, Patil CK, Wodicka L et al. Functional and genomic analyses reveal an essential coordination between the unfolded protein response and ER-associated degradation. Cell 2000; 101(3):249-258.
48. Boisrame A, Kabani M, Beckerich JM et al. Interaction of Kar2p and Sls1p is required for efficient cotranslational translocation of secreted proteins in the yeast Yarrowia lipolytica. J Biol Chem 1998; 273(47):30903-30908.
49. Frydman J. Folding of newly translated proteins in vivo: The role of molecular chaperones. Annu Rev Biochem 2001; 70:603-647.
50. Ahner A, Whyte FM, Brodsky JL. Distinct but overlapping functions of Hsp70, Hsp90, and an Hsp70 nucleotide exchange factor during protein biogenesis in yeast. Arch Biochem Biophys 2005; 435(1):32-41.
51. Li J, Qian X, Sha B. The crystal structure of the yeast Hsp40 Ydj1 complexed with its peptide substrate. Structure (Camb) 2003; 11(12):1475-1483.
52. Lu Z, Cyr DM. Protein folding activity of Hsp70 is modified differentially by the hsp40 co-chaperones Sis1 and Ydj1. J Biol Chem 1998; 273(43):27824-27830.
53. Raynes DA, Graner MW, Bagatell R et al. Increased expression of the Hsp70 co-chaperone HspBP1 in tumors. Tumour Biol 2003; 24(6):281-285.
54. Ciocca DR, Calderwood SK. Heat shock proteins in cancer: Diagnostic, prognostic, predictive, and treatment implications. Cell Stress Chaperones 2005; 10(2):86-103.
55. Harrison CJ, Hayer-Hartl M, Di Liberto M et al. Crystal structure of the nucleotide exchange factor GrpE bound to the ATPase domain of the molecular chaperone DnaK. Science 1997; 276(5311):431-435.

56. Schönfeld HJ, Schmidt D, Schröder H et al. The DnaK chaperone system of Escherichia coli: Quaternary structures and interactions of the DnaK and GrpE components. J Biol Chem 1995; 270(5):2183-2189.
57. Wu B, Wawrzynow A, Zylicz M et al. Structure-function analysis of the Escherichia coli GrpE heat shock protein. EMBO J 1996; 15(18):4806-4816.
58. Flaherty KM, DeLuca-Flaherty C, McKay DB. Three-dimensional structure of the ATPase fragment of a 70K heat-shock cognate protein. Nature 1990; 346(6285):623-628.
59. Briknarova K, Takayama S, Brive L et al. Structural analysis of BAG1 co-chaperone and its interactions with Hsc70 heat shock protein. Nat Struct Biol 2001; 8(4):349-352.
60. Sondermann H, Scheufler C, Schneider C et al. Structure of a Bag/Hsc70 complex: Convergent functional evolution of Hsp70 nucleotide exchange factors. Science 2001; 291(5508):1553-1557.
61. Brive L, Takayama S, Briknarova K et al. The carboxyl-terminal lobe of Hsc70 ATPase domain is sufficient for binding to BAG1. Biochem Biophys Res Commun 2001; 289(5):1099-1105.
62. Vetter IR, Wittinghofer A. The guanine nucleotide-binding switch in three dimensions. Science 2001; 294(5545):1299-1304.
63. Brehmer D, Rüdiger S, Gässler CS et al. Tuning of chaperone activity of Hsp70 proteins by modulation of nucleotide exchange. Nat Struct Biol 2001; 8(5):427-432.
64. Alberti S, Bohse K, Arndt V et al. The Co-chaperone HspBP1 inhibits the CHIP ubiquitin ligase and stimulates the maturation of the cystic fibrosis transmembrane conductance regulator. Mol Biol Cell 2004; 15:4003-4010.
65. Connell P, Ballinger CA, Jiang J et al. The co-chaperone CHIP regulates protein triage decisions mediated by heat-shock proteins. Nat Cell Biol 2001; 3(1):93-96.
66. Meacham GC, Patterson C, Zhang W et al. The Hsc70 co-chaperone CHIP targets immature CFTR for proteasomal degradation. Nat Cell Biol 2001; 3(1):100-105.
67. Grimshaw JPA, Jelesarov I, Sigenthaler RK et al. Thermosensor action of GrpE. J Biol Chem 2003; 278(21):19048-19053.
68. Siegenthaler RK, Christen P. The importance of having thermosensor control in the DnaK chaperone system. J Biol Chem 2005; 280(15):14395-14401.
69. Zhao L, Longo-Guess C, Harris BS et al. Protein accumulation and neurodegeneration in the woozy mutant mouse is caused by disruption of SIL1, a co-chaperone of BiP. Nat Genetics 2005, 37(9):974-979.
70. Kraulis P. MOLSCRIPT: A program to produce both detailed and schematic plots of protein structures. J Appl Cryst 1991; 24:946-950.
71. Merritt EA, Bacon DJ. Raster3D photorealistic graphics. Methods Enzymol 1997; 277:505-524.
72. Gouet P, Courcelle E, Stuart DI et al. ESPript: Multiple sequence alignments in PostScript. Bioinformatics 1999; 15:305-308.

CHAPTER 2

Functions of the Hsp90-Binding FKBP Immunophilins

Marc B. Cox and David F. Smith*

Abstract

Hsp90 functionally interacts with a broad array of client proteins, but in every case examined Hsp90 is accompanied by one or more co-chaperones. One class of co-chaperone contains a tetratricopeptide repeat domain that targets the co-chaperone to the C-terminal region of Hsp90. Within this class are Hsp90-binding peptidylprolyl isomerases, most of which belong to the FK506-binding protein (FKBP) family. Despite the common association of FKBP co-chaperones with Hsp90, it is now clear that the client protein influences, and is influenced by, the particular FKBP bound to Hsp90. Examples include Xap2 in aryl hydrocarbon receptor complexes and FKBP52 in steroid receptor complexes. In this chapter, we discuss the known functional roles played by FKBP co-chaperones and, where possible, relate distinctive functions to structural differences between FKBP members.

Introduction

Immunophilins are a large, functionally diverse group of proteins that are defined by their ability to bind immunosuppressive ligands. The immunophilins minimally contain a peptidyl-prolyl *cis*-trans isomerase (PPIase; also termed rotamase) domain to which the immunosuppressive drugs bind. Early investigations into the PPIase enzymatic activity led to the belief that the immunosuppressive drugs elicited their effects by inhibiting the PPIase activity. However, some compounds binding the PPIase active site efficiently inhibit PPIase activity without inducing immunosuppression, so PPIase activity is not critical for immune responses. It is now known that effector domains on the immunosuppressive drugs project from the PPIase pocket. This allows the immunophilin-drug complex to bind tightly to and inhibit calcineurin or target of rapamycin, signal transduction proteins required for immune responses (see ref. 1 for a detailed review on the mechanisms by which immunophilins and their ligands suppress immune responses).

Since the initial identification of the immunophilin proteins, multiple family members have been identified in all major branches of life. Some immunophilins are small proteins containing only a single PPIase domain while others are large multidomain proteins that contain one or more PPIase domains, as well as additional functional domains. The immunophilins are divided into two groups based on their ability to bind different immunosuppressive ligands: the FK506 binding proteins (FKBP), which also bind rapamycin, and the cyclosporin-A binding proteins or cyclophilins (CyP). The PPIase domains of FKBP and cyclophilins are structurally distinct and likely evolved independently. On the other hand,

*Corresponding Author: David F. Smith—Department of Biochemistry and Molecular Biology, Mayo Clinic Arizona, Scottsdale, Arizona 85259, U.S.A. Email: smith.david26@mayo.edu

Networking of Chaperones by Co-Chaperones, edited by Gregory L. Blatch.

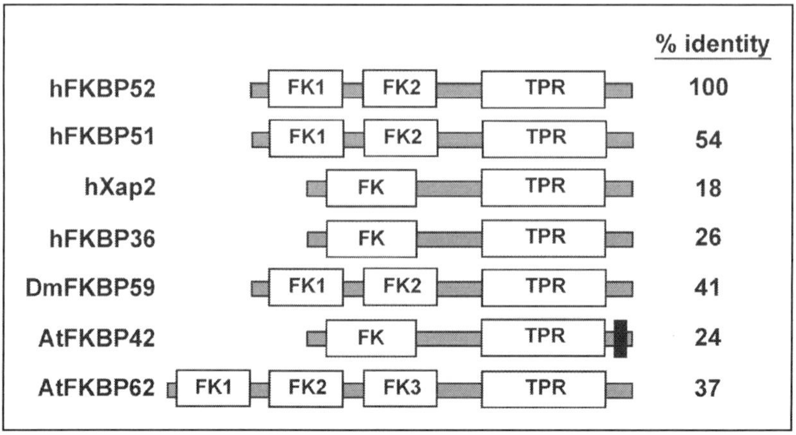

Figure 1. Domain organization of representative Hsp90-binding FKBPs. TPR-containing FKBPs from vertebrate, insect, and plant sources were selected for comparison of domain organizations. The proteins are human FKBP52 (accession number NP_002005), human FKBP51 (acc. # Q13451), human Xap2 (acc. # O00170), human FKBP36 (acc. # NP_003593), *Drosophila melanogaster* FKBP59 (acc. # AAF18387), *Arabadopsis thaliana* FKBP42 (acc. # CAC00654), and *Arabadopsis thaliana* FKBP62 (acc. # AAB82062). The percent amino acid identity of each compared to human FKBP52 was determined from ClustalW alignments (http://www.ebi.ac.uk/clustalw). Each protein shown has at least one FKBP12-like domain (FK), which in some cases has peptidylprolyl isomerase activity and is the binding site for the immunosuppressant drug FK506, and one tetratricopeptide repeat domain (TPR), which is typically an Hsp90 binding site. The black box in the C-terminus of *At*FKBP42 is a transmembrane domain used for anchoring the protein to the plasma and vacuolar membranes.

some members of either the FKBP or cyclophilin families contain a structurally similar tetratricopeptide repeat (TPR) domain that targets binding to heat shock protein 90 (Hsp90).

Hsp90 is an abundant molecular chaperone that interacts with a broad array of protein clients that regulate numerous important cellular pathways. Among the known Hsp90 clients are transcription factors (e.g., steroid hormone receptors, heat shock transcription factor 1, aryl hydrocarbon receptor), both serine/threonine and tyrosine kinases (e.g., Raf and Src-related kinases), and key regulatory enzymes (e.g., nitric oxide synthase and telomerase). A compilation of known Hsp90 clients maintained by Didier Picard at Univ. of Geneva can be accessed at: http://www.picard.ch/downloads/Hsp90interactors.pdf.

In concert with other chaperone proteins, Hsp90 facilitates client folding and proteolytic stability but can also promote client degradation. In the case of steroid receptors, Hsp90 and its associated co-chaperones also regulate receptor activity. Hsp90 binding to steroid receptors must be preceded by transient receptor interactions with Hsp40, Hsp70, and associated co-chaperones. Hsp90, which is recruited as a dimer in the latter stages of complex assembly, binds directly to the receptor ligand binding domain and stabilizes a receptor conformation that is competent for hormone binding. Associated with Hsp90 in the functionally mature receptor complex are p23, a co-chaperone that stabilizes Hsp90 binding to receptor, and any one of several TPR co-chaperones, including the immunophilin/PPIases FKBP52 (also termed p59, Hsp56, p50, HBI, FKBP59, and FKBP4), FKBP51 (also termed p54, FKBP54, and FKBP5), and CyP40, or the protein phosphatase PP5. As discussed below, receptor activity can vary depending on the particular TPR co-chaperone in mature receptor heterocomplexes.

The domain organization for several TPR co-chaperones are compared in Figure 1. These co-chaperones compete for a common binding site in the C-terminal region of Hsp90 that includes the highly conserved -MEEVD sequence that terminates Hsp90. Co-crystallographic structures have shown how an MEEVD pentapeptide associates with the TPR binding pocket.[2,3] Although the TPR domains for each of these co-chaperones are structurally similar and interact in a similar manner with Hsp90, the client protein bound by Hsp90 can influence the rank order of co-chaperone recruitment to Hsp90-client complexes (reviewed in ref. 4). For instance, PP5 and FKBP51 are preferred components in glucocorticoid receptor (GR) complexes, FKBP51 is preferred in progesterone receptor (PR) complexes, and CyP40 is relatively enhanced in estrogen receptor (ER) complexes.[5,6] On the other hand, another TPR-containing FKBP, the hepatitis B virus protein X associated protein 2 (Xap2; also termed AIP, ARA9, and FKBP37) shows little interaction with steroid receptors but is strongly associated with the aryl hydrocarbon receptor-Hsp90 complex.[7,8] The distinctive patterns of preference for co-chaperone association in client complexes is one line of evidence that the co-chaperones bound to Hsp90 can also interact with the Hsp90-bound client.

In addition to FKBP52, FKBP51, and Xap2, several other FKBP family members contain TPR domains that are known or likely to bind Hsp90. FKBP6 (also termed FKBP36) is structurally similar to Xap2 but is required for male fertility and homologous chromosome pairing in meiosis.[9] *Drosophila melanogaster* express a TPR-containing immunophilin (*Dm*FKBP59) that has high similarity to FKBP52/51 in vertebrates.[10,11] Plants have several FKBP genes that encode TPR domains; for example, in *Arabidopsis thaliana* there are 4 such genes: *At*FKBP42, *At*FKBP62, *At*FKBP65 and *At*FKBP72 (reviewed in refs. 12 and 13). Although prokaryotic and Archaeal genomes also contain FKBP family members,[14] none of these genes encode a TPR domain.

Structure/Function Relationships of Steroid Receptor-Associated FKBPs

Three-dimensional crystal structures have been solved for full-length FKBP51 and for overlapping fragments of FKBP52. Both have a similar TPR domain composed of three tandem repeats of the degenerate 34-amino acid motif that characterizes TPR proteins.[15] Each repeat adopts a helix-turn-helix conformation and adjacent units stack in parallel to form a saddle-shaped domain with a concave binding pocket for Hsp90 (Figs. 2A and 2B). In addition to the TPR domain, both FKBP51 and FKBP52 have two N-terminal domains, each of which is structurally similar to FKBP12. FK506-binding and PPIase activities reside in the most N-terminal domain (FK1), which has a pocket and active site residues similar to FKBP12. Due to several amino acid differences, the second domain (FK2) lacks drug binding and PPIase activity;[16] other functions for FK2 have not been identified.

FKBP52 and FKBP51 have distinct functional influences on steroid receptors (see below), but the underlying structure/function relationships are not well understood. FKBP51 and FKBP52 share greater than 60% amino acid sequence similarity, and individual domains do not differ markedly between FKBP51 and FKBP52. The most striking difference in crystal structures relates to apparent domain:domain orientations. The FKBP52 structure shown in Figure 2B is a composite model derived from merging the separate FK1-FK2 and FK2-TPR structures. The composite model suggests that the FKBP52 TPR domain is aligned in a more linear fashion with the FK domains rather than in the kinked conformation seen with FKBP51 (Fig. 2A). In fact, the static orientations shown in crystal structures are likely more dynamic in solution, but the different crystal orientations are perhaps telling. Amino acid side chains unique to FKBP51 form a salt bridge between FK2 and TPR that would stabilize the domain:domain interaction in FKBP51 relative to FKBP52, which lacks this salt bridge. The apparently more flexible structure of FKBP52 might accommodate interactions within the receptor heterocomplex that are constrained in FKBP51.

FKBP51 and FKBP52 also differ in the hinge region connecting FK1 and FK2 domains (FK loop). The FK loop of FKBP52 contains a -TEEED- sequence that has been identified as an in vitro substrate for casein kinase II; the corresponding sequence in FKBP51, -FED-, lacks the threonine phosphorylation site. Phosphorylation of FKBP52 is potentially important since

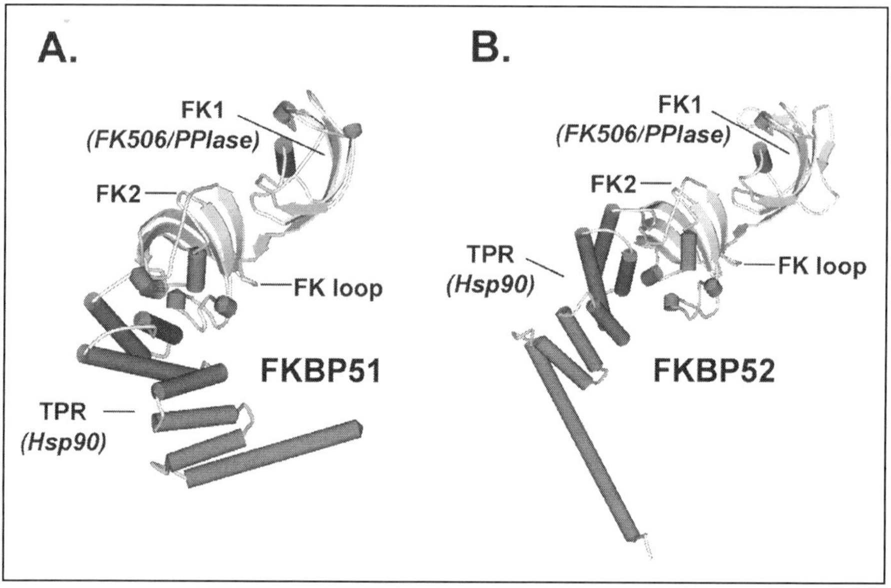

Figure 2. Three-dimensional structures of FKBP51 and FKBP52. This is a depiction of X-ray crystallographic structures for human FKBP51 (A; protein data bank number 1KT0) and a composite of two partial structures for human FKBP52 (B; protein data bank numbers 1Q1C and 1P5Q). In either protein the two FKBP12-like domains are indicated, the first of which has FK506 binding and PPIase activities. A loop structure in the hinge region between FK1 and FK2 is pointed out. The C-terminal TPR domain consists of three helix-loop-helix motifs that form the Hsp90 binding pocket. Structures of the individual domains are highly similar between the two proteins, but the angle between FK2 and TPR domains of FKBP51 is more acute and probably more constrained than in FKBP52. The FKBP51 (A) and FKBP52 (B) structure models shown were constructed using DS ViewerPro version 5.0 (Accelrys Inc., San Diego, CA). A color version of this figure is available online at http://www.eurekah.com.

the phospho-protein is reported to lose Hsp90 binding.[17] Finally, there are a few amino acid differences in and around the respective FK1 PPIase pockets of FKBP52 and FKBP51. Although these FKBPs have similar PPIase activity measured against a model small peptide substrate,[18] FKBP specificity for prolines in the context of a full-length protein might be distinctive. So, for instance, FKBP52 might recognize a receptor proline site that is poorly accessed by FKBP51.

Cellular and Physiological Functions of Hsp90-Associated FKBPs

FKBP52

FKBP52 is expressed in most vertebrate tissues and cell lines, although its expression can be up-regulated by heat stress,[19] by estrogen in MCF-7 breast cancer cells,[20] and by the homeobox transcription factor HoxA-10 in the peri-implantation mouse uterus.[21] FKBP52 associates with steroid receptor complexes in an Hsp90-dependent manner, but FKBP52 is not required in a defined cell-free assembly system for receptor to reach the mature conformation that is competent for hormone binding.[22,23] Nonetheless, FKBP52 in cells potentiates hormone-dependent reporter gene activation by GR,[24] AR,[25] and PR.[26] Potentiation of hormone signaling can be related to an increase in receptor affinity for hormone,[24,27] but there may be additional mechanisms by which FKBP52 enhances receptor activity.

In concordance with hormone binding affinity changes, domain swapping experiments between GR and ER, which is not potentiated by FKBP52, demonstrated that FKBP52 potentiation is localized to the ligand binding domain of GR.[24] FKBP52-dependent potentiation of receptor activity is abrogated in point mutants that are defective for Hsp90 binding or PPIase activity, and potentiation is blocked by the PPIase inhibitor FK506.[24,25] One model to explain these findings is that Hsp90 recruits FKBP52 to the receptor heterocomplex such that the FK1 PPIase can effectively catalyze isomerization of one or more prolines in the receptor ligand binding domain. The resultant change in receptor conformation would translate as an alteration in receptor interactions with hormone or other cellular factors.

FKBP52 has been shown by in vitro studies to have a chaperone activity that is independent of Hsp90-binding or PPIase.[28,29] Like Hsp90 and numerous other chaperone components, FKBP52 can hold misfolded model proteins in a nonaggregated state that is amenable to refolding. The possibility that chaperone holding activity displayed by FKBP52 plays some role in altering receptor activity cannot be dismissed, but this appears unlikely since holding activity is highly redundant among chaperone components. Furthermore, holding activity, unlike FKBP52-dependent potentiation of receptor activity, is neither PPIase- nor Hsp90-dependent. Unfortunately, no one has identified an FKBP52 mutation that disrupts holding activity in a discrete manner.

In an effort to extend biochemical and cellular data on FKBP functions to the physiological level our laboratory has generated FKBP52 gene knockout (52KO) mice. The mutant mice have striking reproductive phenotypes that can be attributed, at least in part, to loss of steroid receptor activity. Male 52KO mice are infertile and display abnormal virilization with persistent nipples, ambiguous external genitalia, and dysgenic seminal vesicles and prostate.[25] These developmental defects are consistent with androgen insensitivity in these tissues. Testicular morphology, descent, histology, and spermatogenesis are normal, and androgen production and release from testes is unimpaired; these developmental features are not highly androgen-dependent. On the other hand, sperm isolated from the epididymis have abnormal tail morphology and reduced motility suggestive of a defect in sperm maturation within the epididymis, a process that is androgen-dependent. Cellular studies confirm that FKBP52 is required for full AR function, which provides a rational explanation for androgen insensitivity in tissues of 52KO males.

52KO females have no gross morphological abnormalities, yet are completely infertile.[26] Oocyte formation and release are not markedly impaired, and oocytes are competent by in vitro and in vivo fertilization. Infertility is due, at least in part, to a maternal failure of embryonic implantation and uterine decidualization. During the early stages of pregnancy, the 52KO uterus does not display the usual molecular or physiological markers for implantation. These events are largely dependent on progesterone actions, and both molecular and cellular studies confirm that FKBP52 is required for full PR activity.

Thus, FKBP52 is critical for reproductive development and success in both male and female mice, and its role can be traced to support of AR and PR function. FKBP52 does not alter ER function in cellular studies and 52KO mice show no signs of estrogen insensitivity. Conversely, results from cellular studies would predict that 52KO mice are resistant to glucocorticoids; consistent with this prediction, 52KO mice have elevated levels of serum corticosterone, although they clearly do not display features suggestive of major glucocorticoid deficits. Endocrine feedback mechanisms may be sufficient to compensate for partial reductions in GR activity, but more in-depth physiological studies are required to determine whether 52KO mice display less apparent immunological or behavioral alterations that could result from partial loss of GR activity.

Apart from the well-established roles of FKBP52 in steroid hormone receptor function, FKBP52, as with other Hsp90 co-chaperones, has been identified in a variety of client-Hsp90 heterocomplexes such as those containing kinases, aryl hydrocarbon receptor, and heat shock transcription factor; however, many of these interactions might reflect passive, transient association of FKBP52 with Hsp90 and have no functional impact on client activity. FKBP52 also has been found to interact directly with the copper transport protein Atox1,[30] which is part of the copper efflux machinery in neurons, interferon regulatory factor 4,[31] which regulates gene expression in B and T lymphocytes, and FKBP associated protein 48,[32] which influences proliferation of Jurkat T cells.[33] A *Drosophila* homolog termed dFKBP59 was found to interact with the Ca^{2+} channel protein TRPL in photoreceptor cells and to influence Ca^{2+} influx.[10] Subsequent studies revealed that FKBP52 similarly interacts with a subset of rat TRPC proteins that form Ca^{2+} channels in the mammalian brain.[34] Each of these interactions were found to be disrupted by FK506 and to target the FKBP52 PPIase domain to specific proline sites in each partner protein. Phenotypes potentially related to these interactions have not yet been assessed in 52KO mice. Not only does FKBP52 interact with proteins, but also FKBP52 is capable of directly binding adeno-associated virus DNA and regulating replication of the viral genome.[35,36] The relevant DNA binding site in FKBP52 has not been identified.

FKBP51

FKBP51/p54/FKBP54 was originally identified as a component of chicken PR complexes[37-39] and is now known to assemble as an Hsp90 co-chaperone with all steroid receptors and other Hsp90-client complexes. FKBP51 is functionally similar in some ways to FKBP52; both have similar PPIase activity in the presence of model peptide substrates, both hold misfolded proteins in a folding competent state, and they compete for binding a common site on Hsp90.[18,40] As noted above, the overall structural similarity of these FKBPs is consistent with these shared functional properties, yet their distinct effects on steroid receptor activity belie these similarities. Another distinction is that the FKBP51 gene is highly inducible by glucocorticoids, androgens and progesterone.[41-47]

FKBP51 acts as an inhibitor of steroid receptor function. The first indication of its inhibitory role came from studies by Scammell and colleagues of glucocorticoid resistance in New World primates.[48,49] In squirrel monkeys GR has a relatively low affinity for hormone yet the cloned monkey GR has an affinity similar to human GR in vitro. This observation led to a search for cellular factors in monkey cells that reduced GR binding affinity. A key factor identified was FKBP51, which is constitutively overexpressed in squirrel monkey cells as well as cells of other New World primates, all of which display some degree of glucocorticoid resistance. Human FKBP51 was also found to inhibit GR function but not to the degree of squirrel monkey FKBP51, which differs in amino acid sequence from its human counterpart at 15 of 457 amino acids. These differences are scattered fairly evenly along the sequence and mapping studies have shown that amino acid changes in several domains contribute to the more potent inhibitory actions of squirrel monkey FKBP51.[50] Crystal structures for both human and squirrel monkey FKBP51 have been solved;[16] although functionally relevant structural changes are not yet apparent, comparison of these structures should ultimately help to understand why inhibitory potencies differ.

In a yeast model for studying functional interactions between steroid receptors and human FKBPs, FKBP51 does not inhibit the activity of GR; however, FKBP51 can effectively reverse the potentiation of GR activity conferred by FKBP52.[24] Therefore, FKBP51 acts as an antagonist of FKBP52, which is unexpected based on measured similarities of FKBP functional properties noted above. FKBP51 has also been shown to inhibit PR function,[46] presumably through a similar inhibition of FKBP52-mediated potentiation. The mechanism by which FKBP51 antagonizes FKBP52's ability to enhance steroid receptor function is not understood. Other Hsp90-binding TPR proteins do not block FKBP52 actions, so it does not appear that competitive displacement of FKBP52 from receptor complexes by FKBP51 can fully account for antagonism. On the other hand, FKBP51 is known to preferentially associate with PR and GR complexes.[6,40] Domain swapping studies indicate that the FK1 PPIase domain partially contributes to antagonism but sequences in the FK2 and TPR domain also play a role.[24,50] Currently, we are genetically mapping sequence differences in FKBP52 and FKBP51 to distinguish how FKBP52 potentiates steroid receptor function and FKBP51 blocks potentiation.

Functional Interactions between FKBP52 and FKBP51

Nuclear Transport

There is some evidence to suggest that FKBP51and FKBP52 have a role in hormone-dependent translocation of GR from the cytosol to nucleus.[51-54] The Pratt laboratory has demonstrated that FKBP52, CyP40 and PP5 bind dynein and therefore link the receptor/Hsp90 complex to the cytoskeleton.[55] Unlike FKBP52, FKBP51 lacks the ability to bind dynein, and hormone-induced switching from FKBP51 to FKBP52 in GR-Hsp90 heterocomplexes has been reported.[52] Thus, one can imagine a model in which FKBP51 holds the receptor in the cytosol until hormone is present, then upon hormone binding FKBP52 replaces FKBP51 in the complex and FKBP52 mediates translocation of the receptor to the nucleus. This is an attractive model, but several observations raise concerns about the general relevance of FKBP52-dynein interactions to steroid receptor function. First, there is a dynamic exchange of Hsp90 co-chaperones within mature Hsp90/steroid receptor complexes before binding hormone,[6] which would seemingly confound receptor localization. Next, although rapid nuclear translocation of GR can be slowed by disrupting cytoskeletal interactions or altering FKBP levels, hormone-bound GR still concentrates in the nucleus within 30 minutes. While one might expect that hormone-dependent activation of a reporter gene might lag in parallel with delayed GR transport, inhibition of reporter expression persists for as long as 16 hours in cells lacking FKBP52. Moreover, FKBP52-dependent potentiation of GR activity is unaltered in yeast that express or lack dynein.[24] Another concern with the transport model is the lack of explanation for why PP5, which binds dynein similar to FKBP52 and assembles preferentially with GR heterocomplexes,[5,55] fails to compensate for loss of FKBP52-dependent GR activity. Clearly, FKBP52 is functioning as more than a linker between GR and dynein complexes. A final consideration relates to differences in subcellular localization of GR, AR, and PR, each of which is similarly potentiated by FKBP52. Although GR is largely localized to the cytosol in the absence of hormone, and thus requires nuclear translocation in response to hormone, AR and PR are more typically localized to the nuclear compartment even in the absence of hormone. It is unlikely that FKBP52 actions relate to nuclear translocation of AR and PR.

Mutual Antagonism

Given that FKBP51 gene expression is inducible by some steroid hormones and FKBP51 can inhibit receptor function, one can reasonably speculate that FKBP51 serves as a cellular modulator of hormone responsiveness. In cells unexposed to hormone, FKBP52 actions would predominate and promote a robust response to hormone; as a consequence, however, FKBP51 levels would rise and partially desensitize cells to a secondary hormone exposure. These effects can be demonstrated in cellular models, but the physiological importance of this mechanism must be established with animal models. Toward this goal, our laboratory has recently generated FKBP51 gene knockout mice. Homozygous mutant animals are grossly normal and reproductively viable (unpublished observations), so FKBP51 does not appear to be critical in the same physiological processes as FKBP52. Nonetheless, modulatory actions of FKBP51 might be relevant but subject to compensatory physiological mechanisms. Interestingly, double knockout of both FKBP51 and FKBP52 genes is embryonic lethal in mice (unpublished observation), suggesting either that FKBP51 and FKBP52 have a critical, mutually redundant function or that FKBP51 and FKBP52 function in a common developmental pathway that requires the distinct actions of both immunophilins.

Aside from its role in steroid receptor complexes, FKBP51 has been shown to regulate NFκB pathways. FKBP51 was identified[56] by a proteomic approach in complex with IKKα, one of the serine/threonine kinases that stimulates phosphorylation and degradation of the NFκB inhibitor IκB. Knockdown of FKBP51 expression was shown to inhibit IKKα activation and thereby block TNFα-induced activation of NFκB, which confirmed the functional significance of FKBP51 in IKKα complexes. Perhaps related to FKBP51-dependent regulation of NFκB pathways, overexpression of FKBP51 has been correlated[57] with idiopathic myelofibrosis, a rare clonal stem cell disorder. Experimental overexpression of FKBP51 was subsequently shown to stimulate NFκB activity and, as a consequence, to increase secretion of pro-fibrotic TGF-β1.[58] IKKα had previously been shown to be an Hsp90 client,[59] so it is possible that, analogous to steroid receptor complexes, FKBP51 assembles with IKKα as a heterocomplex with Hsp90. Whether FKBP51 Hsp90 binding or PPIase is required for regulation of IKKα has not been determined.

Xap2 and FKBP6

Apart from the highly characterized steroid hormone receptor-associated FKBPs, several other TPR-containing FKBPs are present in higher vertebrates. As mentioned in earlier sections of this chapter, Xap2 is a TPR-containing immunophilin that is found almost exclusively in AhR complexes. As the name implies, Xap2 also functionally interacts with the hepatitis B virus protein X.[60] In addition Xap2 is known to have functional interactions with peroxisome proliferator activated receptor α (PPARα),[61] however these interactions have not been extensively characterized. AhR is a ligand-dependent transcription factor that mediates the physiological response to specific environmental contaminants termed polycyclic aromatic hydrocarbons, the most notorious of which is 2,3,7,8-tetrachlorodibenzo-p-dioxin. Similar to steroid receptors, AhR requires assembly with Hsp90 and p23 to achieve a mature ligand-binding conformation (reviewed in ref. 62), although the AhR ligand binding domain is unrelated to steroid receptor ligand binding domains. AhR complexes also contain an FKBP component, but in this case it is Xap2 rather than FKBP52 or FKBP51.

As with FKBP51 and FKBP52, Xap2 has a C-terminal TPR domain that is known to facilitate binding to the MEEVD motif on Hsp90 (Fig. 1).[63] In addition Xap2 contains one N-terminal FK domain that lacks drug binding and also likely lacks PPIase activity. Although the FK domain is not required for Hsp90 binding, it is required for an interaction with the AhR-Hsp90 complex that functionally influences receptor activity.[63,64] In a cell-free assembly system that lacks Xap2, AhR is capable of assembling with Hsp90 and binding ligand, and upon ligand binding AhR is capable of binding AhR response elements on DNA.[8] Again, similar to FKBP52 or FKBP51 in steroid receptor complexes, Xap2 is not required for basal maturation of AhR activity, but in both yeast and mammalian systems, Xap2 can modulate AhR-mediated reporter gene expression.[7,8,63,65] By titrating the relative level of Xap2 protein in cells, AhR activity can be enhanced or decreased. For example, when Xap2 is expressed at a level 2- to 3-fold higher than normal, binding of p23 in the AhR-Hsp90 complex is reduced.[66] Displacement of p23 by high levels of Xap2 would destabilize binding of Hsp90 to AhR and reduce the proportion of AhR in functionally mature complexes. Conversely, there is also evidence that at elevated Xap2 levels AhR is protected from ubiquitination and proteosomal degradation which would increase total AhR levels.[67-71] Finally, several studies suggest that Xap2 facilitates nucleocytoplasmic shuttling of AhR following ligand binding.[71-75]

The physiological relevance of Xap2 interactions with AhR complexes has not been examined in a whole animal model, but Xap2 could potentially influence any of several physiological and pathological pathways mediated by AhR. Mice that are homozygous for a disrupted AhR gene have many physiological and developmental defects; among these are immune system impairment, hepatic fibrosis, cardiac hypertrophy, impaired insulin regulation, and defects in ovarian and vascular development.[76-79] In addition, many of the toxic and teratogenic effects produced by AhR ligands require an intact AhR signaling pathway.[80,81] For example, dioxin induced defects in prostate development are absent in AhR knockout mice.[82] Development of a mouse strain lacking Xap2 could be very helpful in determining the role Xap2 plays in these processes and might validate Xap2 as a potential target for therapeutic intervention.

In addition to Xap2, vertebrates contain FKBP36 (gene name FKBP6 in humans), another TPR-containing FKBP that is structurally similar to Xap2 yet functionally distinct. FKBP36 has a single N-terminal FK domain and a C-terminal TPR domain. In vitro studies show that

FKBP36 binds Hsp90 and can assemble with steroid receptor complexes (unpublished observation), but there is currently no evidence that FKBP36 alters receptor activity. FKBP36 mRNA is broadly expressed in vertebrate tissues with an exceptionally high level observed in the testis; male FKBP6 knockout mice lack sperm and FKBP36 was shown to be a critical component in meiotic synaptonemal complexes.[9] Patients with Williams syndrome, which is characterized by congenital cardiovascular defects, dysmorphic facial features, mental retardation, growth defects, azoospermia, and hypercalcemia, are typically haploinsufficient for FKBP6;[83] however, the contribution of FKBP6 deletion in this syndrome is not clear since several contiguous genes on chromosome 11, including genes for elastin and LIM-Kinase 1, are also deleted in these patients and clearly contribute to some phenotypic aspects.

Plant FKBPs

Hsp90-binding TPR immunophilins have been identified in all eukaryotes examined. A few examples of plant TPR-containing FKBPs are shown in Figure 1. The TPR domain of each FKBP is very similar in amino acid sequence to that of vertebrate proteins; these are presumed to bind Hsp90, but that has not been determined in all cases. The plant and insect FKBPs contain one or more PPIase-related domain and can contain other functional domains. For example, *At*FKBP42 contains a C-terminal transmembrane domain that localizes the protein to the inner plasma membrane and the vacuolar membrane.[84-86]

There is ample evidence to suggest that the plant and insect FKBPs are physiologically important. Mutations in *AtFKBP42* cause the severe developmental phenotypes termed *twisted dwarf 1* (*TWD*)[85] and *ultracurvata* (*UCU2*).[87] The mechanism by which these phenotypes occur likely involves impairment of membrane transport as *At*FKBP42 is known to interact with several ATP-binding cassette transporters on the plasma and vacuolar membranes.[85,86,88] Mutations in *AtFKBP72* result in a class of mutants termed *pasticcino* or *pas* mutants, which are characterized by a wide variety of developmental defects.[89] Two Hsp90-binding TPR FKBPs in wheat, wFKBP72 and the heat shock-inducible wFKBP77, have been shown in transgenic plants to distinctively influence developmental patterns.[90]

Summary

In addressing the physiological importance of PPIases, Heitman and colleagues[91] generated an *S. cerevisiae* strain that lacked all 12 PPIase genes in the FKBP and cyclophilin families; the pluri-mutant strain displayed some growth abnormalities but was viable, thus demonstrating that these genes collectively are nonessential in yeast. Nonetheless, it has become increasingly clear that the Hsp90-binding PPIases, through interactions with steroid receptors, kinases, and other cellular factors, play important physiological and potentially pathological roles in mammals. Elucidation of these roles, definition of underlying molecular mechanisms, and identification of specific inhibitors will likely quicken in the coming few years and lead to therapeutic targeting of individual PPIases.

Acknowledgements

Studies in the authors' laboratory were supported by NIH R01-DK48218 and the Mayo Foundation.

References

1. Hamilton GS, Steiner JP. Immunophilins: Beyond immunosuppression. J Med Chem 1998; 41(26):5119-5143.
2. Scheufler C, Brinker A, Bourenkov G et al. Structure of TPR domain-peptide complexes: Critical elements in the assembly of the Hsp70-Hsp90 multichaperone machine. Cell 2000; 101(2):199-210.
3. Wu B, Li P, Liu Y et al. 3D structure of human FK506-binding protein 52: Implications for the assembly of the glucocorticoid receptor/Hsp90/immunophilin heterocomplex. Proc Natl Acad Sci USA 2004; 101(22):8348-8353.

4. Riggs DL, Cox MB, Cheung-Flynn J et al. Functional specificity of co-chaperone interactions with Hsp90 client proteins. Crit Rev Biochem Mol Biol 2004; 39:279-295.
5. Silverstein AM, Galigniana MD, Chen MS et al. Protein phosphatase 5 is a major component of glucocorticoid receptor.hsp90 complexes with properties of an FK506-binding immunophilin. J Biol Chem 1997; 272(26):16224-16230.
6. Barent RL, Nair SC, Carr DC et al. Analysis of FKBP51/FKBP52 chimeras and mutants for Hsp90 binding and association with progesterone receptor complexes. Mol Endocrinol 1998; 12(3):342-354.
7. Ma Q, Whitlock Jr JP. A novel cytoplasmic protein that interacts with the Ah receptor, contains tetratricopeptide repeat motifs, and augments the transcriptional response to 2,3,7,8-tetrachlorodibenzo-p-dioxin. J Biol Chem 1997; 272(14):8878-8884.
8. Meyer BK, Pray-Grant MG, Vanden Heuvel JP et al. Hepatitis B virus X-associated protein 2 is a subunit of the unliganded aryl hydrocarbon receptor core complex and exhibits transcriptional enhancer activity. Mol Cell Biol 1998; 18(2):978-988.
9. Crackower MA, Kolas NK, Noguchi J et al. Essential role of Fkbp6 in male fertility and homologous chromosome pairing in meiosis. Science 2003; 300(5623):1291-1295.
10. Goel M, Garcia R, Estacion M et al. Regulation of drosophila trpl channels by immunophilin fkbp59. J Biol Chem 2001; 276(42):38762-38773.
11. Zaffran S. Molecular cloning and embryonic expression of dFKBP59, a novel Drosophila FK506-binding protein. Gene 2000; 246(1-2):103-109.
12. Romano P, Gray J, Horton P et al. Plant immunophilins: Functional versatility beyond protein maturation. New Phytol 2005; 166(3):753-769.
13. He Z, Li L, Luan S. Immunophilins and parvulins. Superfamily of peptidyl prolyl isomerases in Arabidopsis. Plant Physiol 2004; 134(4):1248-1267.
14. Maruyama T, Suzuki R, Furutani M. Archaeal peptidyl prolyl cis-trans isomerases (PPIases) update 2004. Front Biosci 2004; 9:1680-1720.
15. Blatch GL, Lassle M. The tetratricopeptide repeat: A structural motif mediating protein-protein interactions. Bioessays 1999; 21(11):932-939.
16. Sinars CR, Cheung-Flynn J, Rimerman RA et al. Structure of the large FK506-binding protein FKBP51, an Hsp90-binding protein and a component of steroid receptor complexes. Proc Natl Acad Sci USA 2003; 100(3):868-873.
17. Miyata Y, Chambraud B, Radanyi C et al. Phosphorylation of the immunosuppressant FK506-binding protein FKBP52 by casein kinase II: Regulation of HSP90-binding activity of FKBP52. Proc Natl Acad Sci USA 1997; 94(26):14500-14505.
18. Pirkl F, Buchner J. Functional analysis of the Hsp90-associated human peptidyl prolyl cis/trans isomerases FKBP51, FKBP52 and Cyp40. J Mol Biol 2001; 308(4):795-806.
19. Sanchez ER. Hsp56: A novel heat shock protein associated with untransformed steroid receptor complexes. J Biol Chem 1990; 265(36):22067-22070.
20. Kumar P, Mark PJ, Ward BK et al. Estradiol-regulated expression of the immunophilins cyclophilin 40 and FKBP52 in MCF-7 breast cancer cells. Biochem Biophys Res Commun 2001; 284(1):219-225.
21. Daikoku T, Tranguch S, Friedman DB et al. Proteomic analysis identifies immunophilin FK506 binding protein 4 (FKBP52) as a downstream target of Hoxa10 in the periimplantation mouse uterus. Mol Endocrinol 2005; 19(3):683-697.
22. Dittmar KD, Hutchison KA, Owens-Grillo JK et al. Reconstitution of the steroid receptor.hsp90 heterocomplex assembly system of rabbit reticulocyte lysate. J Biol Chem 1996; 271(22):12833-12839.
23. Kosano H, Stensgard B, Charlesworth MC et al. The assembly of progesterone receptor-hsp90 complexes using purified proteins. J Biol Chem 1998; 273(49):32973-32979.
24. Riggs DL, Roberts PJ, Chirillo SC et al. The Hsp90-binding peptidylprolyl isomerase FKBP52 potentiates glucocorticoid signaling in vivo. EMBO J 2003; 22(5):1158-1167.
25. Cheung-Flynn J, Prapapanich V, Cox MB et al. Physiological role for the co-chaperone FKBP52 in androgen receptor signaling. Mol Endocrinol 2005; 19(6):1654-1666.
26. Tranguch S, Cheung-Flynn J, Daikoku T et al. Co-chaperone immunophilin FKBP52 is critical to uterine receptivity for embryo implantation. Proc Natl Acad Sci USA 2005; 102(40):14326-14331.
27. Davies TH, Ning YM, Sanchez ER. Differential control of glucocorticoid receptor hormone-binding function by tetratricopeptide repeat (TPR) proteins and the immunosuppressive ligand FK506. Biochemistry 2005; 44(6):2030-2038.
28. Bose S, Weikl T, Bugl H et al. Chaperone function of Hsp90-associated proteins. Science 1996; 274(5293):1715-1717.
29. Pirkl F, Fischer E, Modrow S et al. Localization of the chaperone domain of FKBP52. J Biol Chem 2001; 276(40):37034-37041.

30. Sanokawa-Akakura R, Dai H, Akakura S et al. A novel role for the immunophilin FKBP52 in copper transport. J Biol Chem 2004; 279(27):27845-27848.
31. Mamane Y, Sharma S, Petropoulos L et al. Posttranslational regulation of IRF-4 activity by the immunophilin FKBP52. Immunity 2000; 12(2):129-140.
32. Chambraud B, Radanyi C, Camonis JH et al. FAP48, a new protein that forms specific complexes with both immunophilins FKBP59 and FKBP12. Prevention by the immunosuppressant drugs FK506 and rapamycin. J Biol Chem 1996; 271(51):32923-32929.
33. Krummrei U, Baulieu EE, Chambraud B. The FKBP-associated protein FAP48 is an antiproliferative molecule and a player in T cell activation that increases IL2 synthesis. Proc Natl Acad Sci USA 2003; 100(5):2444-2449.
34. Sinkins WG, Goel M, Estacion M et al. Association of immunophilins with mammalian TRPC channels. J Biol Chem 2004; 279(33):34521-34529.
35. Qing K, Hansen J, Weigel-Kelley KA et al. Adeno-associated virus type 2-mediated gene transfer: Role of cellular FKBP52 protein in transgene expression. J Virol 2001; 75(19):8968-8976.
36. Zhong L, Li W, Yang Z et al. Improved transduction of primary murine hepatocytes by recombinant adeno-associated virus 2 vectors in vivo. Gene Ther 2004; 11(14):1165-1169.
37. Smith DF, Faber LE, Toft DO. Purification of unactivated progesterone receptor and identification of novel receptor-associated proteins. J Biol Chem 1990; 265(7):3996-4003.
38. Smith DF, Baggenstoss BA, Marion TN et al. Two FKBP-related proteins are associated with progesterone receptor complexes. J Biol Chem 1993; 268(24):18365-18371.
39. Smith DF, Albers MW, Schreiber SL et al. FKBP54, a novel FK506-binding protein in avian progesterone receptor complexes and HeLa extracts. J Biol Chem 1993; 268(32):24270-24273.
40. Nair SC, Rimerman RA, Toran EJ et al. Molecular cloning of human FKBP51 and comparisons of immunophilin interactions with Hsp90 and progesterone receptor. Mol Cell Biol 1997; 17(2):594-603.
41. Baughman G, Wiederrecht GJ, Campbell NF et al. FKBP51, a novel T-cell-specific immunophilin capable of calcineurin inhibition. Mol Cell Biol 1995; 15(8):4395-4402.
42. Kester HA, van der Leede BM, van der Saag PT et al. Novel progesterone target genes identified by an improved differential display technique suggest that progestin-induced growth inhibition of breast cancer cells coincides with enhancement of differentiation. J Biol Chem 1997; 272(26):16637-16643.
43. Zhu W, Zhang JS, Young CY. Silymarin inhibits function of the androgen receptor by reducing nuclear localization of the receptor in the human prostate cancer cell line LNCaP. Carcinogenesis 2001; 22(9):1399-1403.
44. Yoshida NL, Miyashita T, UM et al. Analysis of gene expression patterns during glucocorticoid-induced apoptosis using oligonucleotide arrays. Biochem Biophys Res Commun 2002; 293(4):1254-1261.
45. Vermeer H, Hendriks-Stegeman BI, van der Burg B et al. Glucocorticoid-induced increase in lymphocytic FKBP51 messenger ribonucleic acid expression: A potential marker for glucocorticoid sensitivity, potency, and bioavailability. J Clin Endocrinol Metab 2003; 88(1):277-284.
46. Hubler TR, Denny WB, Valentine DL et al. The FK506-binding immunophilin FKBP51 is transcriptionally regulated by progestin and attenuates progestin responsiveness. Endocrinology 2003; 144(6):2380-2387.
47. Febbo PG, Lowenberg M, Thorner AR et al. Androgen mediated regulation and functional implications of fkbp51 expression in prostate cancer. J Urol 2005; 173(5):1772-1777.
48. Reynolds PD, Ruan Y, Smith DF et al. Glucocorticoid resistance in the squirrel monkey is associated with overexpression of the immunophilin FKBP51. J Clin Endocrinol Metab 1999; 84(2):663-669.
49. Denny WB, Valentine DL, Reynolds PD et al. Squirrel monkey immunophilin FKBP51 is a potent inhibitor of glucocorticoid receptor binding. Endocrinology 2000; 141(11):4107-4113.
50. Denny WB, Prapapanich V, Smith DF et al. Structurefunction analysis of squirrel monkey FK506-binding protein 51, a potent inhibitor of glucocorticoid receptor activity. Endocrinology 2005; 146(7):3194-3201.
51. Czar MJ, Lyons RH, Welsh MJ et al. Evidence that the FK506-binding immunophilin heat shock protein 56 is required for trafficking of the glucocorticoid receptor from the cytoplasm to the nucleus. Mol Endocrinol 1995; 9(11):1549-1560.
52. Davies TH, Ning YM, Sanchez ER. A new first step in activation of steroid receptors: Hormone-induced switching of FKBP51 and FKBP52 immunophilins. J Biol Chem 2002; 277(7):4597-4600.

53. Wochnik GM, Ruegg J, Abel GA et al. FK506-binding proteins 51 and 52 differentially regulate dynein interaction and nuclear translocation of the glucocorticoid receptor in mammalian cells. J Biol Chem 2005; 280(6):4609-4616.
54. Pratt WB, Galigniana MD, Harrell JM et al. Role of hsp90 and the hsp90-binding immunophilins in signalling protein movement. Cell Signal 2004; 16(8):857-872.
55. Galigniana MD, Harrell JM, Murphy PJ et al. Binding of hsp90-associated immunophilins to cytoplasmic dynein: Direct binding and in vivo evidence that the peptidylprolyl isomerase domain is a dynein interaction domain. Biochemistry 2002; 41(46):13602-13610.
56. Bouwmeester T, Bauch A, Ruffner H et al. A physical and functional map of the human TNF-alpha/NF-kappa B signal transduction pathway. Nat Cell Biol 2004; 6(2):97-105.
57. Giraudier S, Chagraoui H, Komura E et al. Overexpression of FKBP51 in idiopathic myelofibrosis regulates the growth factor independence of megakaryocyte progenitors. Blood 2002; 100(8):2932-2940.
58. Komura E, Tonetti C, Penard-Lacronique V et al. Role for the nuclear factor kappaB pathway in transforming growth factor-beta1 production in idiopathic myelofibrosis: Possible relationship with FK506 binding protein 51 overexpression. Cancer Res 2005; 65(8):3281-3289.
59. Broemer M, Krappmann D, Scheidereit C. Requirement of Hsp90 activity for IkappaB kinase (IKK) biosynthesis and for constitutive and inducible IKK and NF-kappaB activation. Oncogene 2004; 23(31):5378-5386.
60. Kuzhandaivelu N, Cong YS, Inouye C et al. XAP2, a novel hepatitis B virus X-associated protein that inhibits X transactivation. Nucleic Acids Res 1996; 24(23):4741-4750.
61. Sumanasekera WK, Tien ES, Turpey R et al. Evidence that peroxisome proliferator-activated receptor alpha is complexed with the 90-kDa heat shock protein and the hepatitis virus B X-associated protein 2. J Biol Chem 2003; 278(7):4467-4473.
62. Petrulis JR, Perdew GH. The role of chaperone proteins in the aryl hydrocarbon receptor core complex. Chem Biol Interact 2002; 141(1-2):25-40.
63. Carver LA, LaPres JJ, Jain S et al. Characterization of the Ah receptor-associated protein, ARA9. J Biol Chem 1998; 273(50):33580-33587.
64. Kazlauskas A, Poellinger L, Pongratz I. Two distinct regions of the immunophilin-like protein XAP2 regulate dioxin receptor function and interaction with hsp90. J Biol Chem 2002; 277(14):11795-11801.
65. Miller CA. Two tetratricopeptide repeat proteins facilitate human aryl hydrocarbon receptor signalling in yeast. Cell Signal 2002; 14(7):615-623.
66. Hollingshead BD, Petrulis JR, Perdew GH. The aryl hydrocarbon (Ah) receptor transcriptional regulator hepatitis B virus X-associated protein 2 antagonizes p23 binding to Ah receptor-Hsp90 complexes and is dispensable for receptor function. J Biol Chem 2004; 279(44):45652-45661.
67. Lees MJ, Peet DJ, Whitelaw ML. Defining the role for XAP2 in stabilization of the dioxin receptor. J Biol Chem 2003; 278(38):35878-35888.
68. LaPres JJ, Glover E, Dunham EE et al. ARA9 modifies agonist signaling through an increase in cytosolic aryl hydrocarbon receptor. J Biol Chem 2000; 275(9):6153-6159.
69. Meyer BK, Petrulis JR, Perdew GH. Aryl hydrocarbon (Ah) receptor levels are selectively modulated by hsp90-associated immunophilin homolog XAP2. Cell Stress Chaperones 2000; 5(3):243-254.
70. Meyer BK, Perdew GH. Characterization of the AhR-hsp90-XAP2 core complex and the role of the immunophilin-related protein XAP2 in AhR stabilization. Biochemistry 1999; 38(28):8907-8917.
71. Kazlauskas A, Poellinger L, Pongratz I. The immunophilin-like protein XAP2 regulates ubiquitination and subcellular localization of the dioxin receptor. J Biol Chem 2000; 275(52):41317-41324.
72. Berg P, Pongratz I. Two parallel pathways mediate cytoplasmic localization of the dioxin (aryl hydrocarbon) receptor. J Biol Chem 2002; 277(35):32310-32319.
73. Petrulis JR, Hord NG, Perdew GH. Subcellular localization of the aryl hydrocarbon receptor is modulated by the immunophilin homolog hepatitis B virus X-associated protein 2. J Biol Chem 2000; 275(48):37448-37453.
74. Kazlauskas A, Sundstrom S, Poellinger L et al. The hsp90 chaperone complex regulates intracellular localization of the dioxin receptor. Mol Cell Biol 2001; 21(7):2594-2607.
75. Petrulis JR, Kusnadi A, Ramadoss P et al. The hsp90 Co-chaperone XAP2 alters importin beta recognition of the bipartite nuclear localization signal of the Ah receptor and represses transcriptional activity. J Biol Chem 2003; 278(4):2677-2685.
76. Fernandez-Salguero P, Pineau T, Hilbert DM et al. Immune system impairment and hepatic fibrosis in mice lacking the dioxin-binding Ah receptor. Science 1995; 268(5211):722-726.
77. Lahvis GP, Pyzalski RW, Glover E et al. The aryl hydrocarbon receptor is required for developmental closure of the ductus venosus in the neonatal mouse. Mol Pharmacol 2005; 67(3):714-720.

78. Thackaberry EA, Bedrick EJ, Goens MB et al. Insulin regulation in AhR-null mice: Embryonic cardiac enlargement, neonatal macrosomia, and altered insulin regulation and response in pregnant and aging AhR-null females. Toxicol Sci 2003; 76(2):407-417.
79. Benedict JC, Lin TM, Loeffler IK et al. Physiological role of the aryl hydrocarbon receptor in mouse ovary development. Toxicol Sci 2000; 56(2):382-388.
80. Mimura J, Fujii Kuriyama Y. Functional role of AhR in the expression of toxic effects by TCDD. Biochim Biophys Acta 2003; 1619(3):263-268.
81. Fernandez-Salguero PM, Hilbert DM, Rudikoff S et al. Aryl-hydrocarbon receptor-deficient mice are resistant to 2,3,7,8-tetrachlorodibenzo-p-dioxin-induced toxicity. Toxicol Appl Pharmacol 1996; 140(1):173-179.
82. Lin TM, Ko K, Moore RW et al. Effects of aryl hydrocarbon receptor null mutation and in utero and lactational 2,3,7,8-tetrachlorodibenzo-p-dioxin exposure on prostate and seminal vesicle development in C57BL/6 mice. Toxicol Sci 2002; 68(2):479-487.
83. Meng X, Lu X, Morris CA et al. A novel human gene FKBP6 is deleted in Williams syndrome. Genomics 1998; 52(2):130-137.
84. Kamphausen T, Fanghanel J, Neumann D et al. Characterization of Arabidopsis thaliana AtFKBP42 that is membrane-bound and interacts with Hsp90. Plant J 2002; 32(3):263-276.
85. Geisler M, Kolukisaoglu HU, Bouchard R et al. TWISTED DWARF1, a unique plasma membrane-anchored immunophilin-like protein, interacts with Arabidopsis multidrug resistance-like transporters AtPGP1 and AtPGP19. Mol Biol Cell 2003; 14(10):4238-4249.
86. Geisler M, Girin M, Brandt S et al. Arabidopsis immunophilin-like TWD1 functionally interacts with vacuolar ABC transporters. Mol Biol Cell 2004; 15(7):3393-3405.
87. Perez-Perez JM, Ponce MR, Micol JL. The ULTRACURVATA2 gene of Arabidopsis encodes an FK506-binding protein involved in auxin and brassinosteroid signaling. Plant Physiol 2004; 134(1):101-117.
88. Liu G, Sanchez-Fernandez R, Li ZS et al. Enhanced multispecificity of arabidopsis vacuolar multidrug resistance-associated protein-type ATP-binding cassette transporter, AtMRP2. J Biol Chem 2001; 276(12):8648-8656.
89. Vittorioso P, Cowling R, Faure JD et al. Mutation in the Arabidopsis PASTICCINO1 gene, which encodes a new FK506- binding protein-like protein, has a dramatic effect on plant development. Mol Cell Biol 1998; 18(5):3034-3043.
90. Kurek I, Stoger E, Dulberger R et al. Overexpression of the wheat FK506-binding protein 73 (FKBP73) and the heat-induced wheat FKBP77 in transgenic wheat reveals different functions of the two isoforms. Transgenic Res 2002; 11(4):373-379.
91. Dolinski K, Muir S, Cardenas M et al. All cyclophilins and FK506 binding proteins are, individually and collectively, dispensable for viability in Saccharomyces cerevisiae. Proc Natl Acad Sci USA 1997; 94(24):13093-13098.

Chapter 3

Hop:
An Hsp70/Hsp90 Co-Chaperone That Functions Within and Beyond Hsp70/Hsp90 Protein Folding Pathways

Sheril Daniel, Csaba Söti, Peter Csermely, Graeme Bradley* and Gregory L. Blatch*

Abstract

Molecular chaperones and their co-chaperones are crucial for the facilitation of efficient protein folding, and prevention of denaturation and aggregation of nascent polypeptides. Hsp70/Hsp90 organizing protein (Hop), a co-chaperone of the two major molecular chaperones, heat shock protein 70 (Hsp70) and heat shock protein 90 (Hsp90), facilitates their interaction by acting as an adaptor between the two chaperones, so that substrate is efficiently transferred from Hsp70 to Hsp90. Although initial studies reported its scaffolding properties to be its primary function, recent findings suggest an additional modulatory effect of Hop on the activities of Hsp70 and Hsp90. In addition, a more diverse role of Hop, involving structurally and functionally unrelated biomolecules and complexes, is currently being revealed. This review focuses on the integratory and modulatory effects of Hop on the Hsp70 and Hsp90 protein folding pathways, and puts forward evidence and theories regarding its multifaceted roles within various biological systems.

Introduction

The efficient folding of polypeptides is extremely challenging within the complex cellular environment due to various reasons, including proteotoxic conditions such as heat stress, anoxia, exposure to heavy metals or other chemical agents. The assistance of molecular chaperones, a group of proteins that are adapted to facilitate protein folding, has thus proven to be critical in this regard. Molecular chaperones are known to interact reversibly with nascent polypeptide chains in an attempt to reduce inappropriate interactions that can otherwise lead to poorly reversible conformations and aggregations.[1] Heat shock proteins (Hsps) are a group of cytoprotective proteins synthesized in response to various kinds of cell stress, and they form the central components of the molecular chaperone machinery.[2] They protect functional proteins from irreversible denaturation as well as assist them in renaturation. Two of the most studied heat shock protein families are Hsp70, a structurally conserved protein with a role in the survival of the organism and Hsp90, one of the most abundant cytosolic proteins in eukaryotes, essential for its viability.[3]

Hsp70, found in eukaryotes, eubacteria and many archaea, is primarily involved in protecting proteins against misfolding and aggregation within the cell's overcrowded environment.[4] It

*Corresponding Authors: Gregory L. Blatch and Graeme Bradley—Department of Biochemistry, Microbiology and Biotechnology, Rhodes University, Grahamstown 6140, South Africa. Emails: g.blatch@ru.ac.za and g.bradley@ru.ac.za

Networking of Chaperones by Co-Chaperones, edited by Gregory L. Blatch.

is well known for its role in recognizing, binding and stabilizing unfolded proteins, translocation of newly synthesized proteins, protein degradation and protection of the cell against the effects of cellular stress.[5-7] The Hsp70 family is composed of four major members: cytosolic constitutive heat shock cognate 70 (Hsc70), cytosolic inducible heat shock protein 70 (Hsp70), endoplasmic reticulum Hsp70 which is also known as the immunoglobulin heavy-chain binding protein (BiP), and mitochondrial Hsp70 (mtHsp70).[8]

The chaperone activity of Hsp70 is regulated by co-chaperones, which usually act by modifying the ATPase cycle of Hsp70. Hsp70 in the ATP bound state has a lower affinity for substrates than the ADP bound state. ATP hydrolysis converts Hsp70 to its higher affinity substrate-binding state and subsequent nucleotide exchange allows for substrate release and return of Hsp70 to its lower affinity state.[9] Hsp40 proteins are well-known co-chaperones of Hsp70, regulating the activity of Hsp70 by stimulating its ATP hydrolysis step (see Chapter by Rosser and Cyr). GrpE is an additional co-chaperone of prokaryotic Hsp70 that acts by stimulating nucleotide exchange thereby enhancing the basal ATPase activity of Hsp70 up to 50 times.[10] A functional equivalent of GrpE in eukaryotes, Bag-1 (Bcl2-associated anthanogene), also regulates Hsp70 nucleotide exchange in a similar manner to GrpE (see Chapter by Brodsky and Bracher).[11] In eukaroyotes, Hsp70-interacting protein (Hip) stimulates the assembly of the Hsp70-Hsp40-substrate complex and stabilizes the ADP-bound form of Hsp70 so that the unfolded polypeptide has more time to attain its proper conformation before being released from the chaperone complex. Once released, the polypeptide either folds to its native state, or is passed on to other molecular chaperones, which include the Hsp90 chaperone machinery.[12]

Hsp90 is a ubiquitous and abundant cytosolic molecular chaperone that is conserved from bacteria to mammals.[13,14] It plays a variety of roles in processes such as protein restoration, protein degradation, signaling, cytoplasmic organization, nuclear transport, DNA rearrangements, DNA-protein interactions, the cell cycle and apoptosis. Hsp90 interacts with a diverse range of proteins (referred to as client and/or substrate proteins) and ensures the folding and maturation of these molecules, which includes steroid receptors, phosphatases, protein kinases and other signaling intermediates of the mitogenic signal transduction pathway.[15,16] In vitro experiments have shown binding and anti-aggregation properties of purified Hsp90 to denatured protein, however a cohort of co-chaperones, which form several subcomplexes with Hsp90, are necessary for it to carry out its functions in vivo. These co-chaperones enable Hsp90 to attend to such a versatile range of client proteins.[17,18] Some of these co-chaperones also interact directly with Hsp90 substrates, as well as display chaperone activity on their own.[19]

Hsp90 contains two ATP binding sites within the N and C- terminal domains and ATP hydrolysis is of crucial importance for Hsp90 functioning in vivo.[17,20-23] The ADP-bound form of Hsp90 is described as "relaxed" and therefore ideal for client protein loading, whereas the ATP-bound form of Hsp90 is described as a "closed" conformation, which is capable of tightly retaining the substrate.[22] Conversion of the ATP state of Hsp90 to its ADP form allows for the efficient release of substrate,[21,24] and this is in stark contrast to the ATP regulated substrate-binding cycle of Hsp70. The benzoquinone ansamycin antibiotic geldanamycin blocks this cycle by maintaining Hsp90 in an ADP-bound state, thereby acting as a specific inhibitor of Hsp90.[17] The Hch1/Aha1 proteins have been identified as Hsp90 co-chaperones, accelerating the ATPase activity of yeast Hsp90 to 12 times its basal level.[25] Cdc37 (p50) is an inhibitor of the Hsp90 ATPase activity, and this suppression is restored to normal levels when Cdc37/p50 is displaced by Cpr6.[24]

Although both Hsp70 and Hsp90 protein folding systems act independently of each other and on different substrates, some protein substrates are processed by Hsp70 and then transferred to Hsp90. The collaboration between the two major chaperone machineries, Hsp70 and Hsp90, is coordinated by a number of co-chaperones. This review will focus on Hop, the Hsp70/Hsp90 organizing protein, which is a unique co-chaperone that interacts with both Hsp70 and Hsp90, bringing them together in a molecular chaperone complex.

Hop (Hsp70/Hsp90 Organizing Protein)

The 60-kDa protein Hop, was first identified by Nicolet and Craig[26] during a genetic screen for proteins that were involved in the heat shock response in yeast. Hop has been found to associate with Hsp70 and Hsp90 within intermediate steroid receptor complexes and appears to be essential for the in vitro assembly of steroid receptors with Hsp90.[27,28] Homologues of Hop have also been identified in humans,[29] mice,[30] rats,[31] insects,[32] plants,[33] and parasites[34] and are classified as belonging to the stress-inducible protein 1 (STI1) family.[26] In this review, Hop will refer to the protein of mammalian origin, and that of a specific species such as yeast or mouse, will be designated with the first letter of the species next to Hop, eg. yHop (yeast Hop) and mHop (mouse Hop).

The presence of nine tetratricopeptide repeat (TPR) motifs structurally defines homologues of Hop, in which the TPR motifs are grouped into three domains, each comprising three TPRs (Fig. 1A). TPRs are protein-protein interaction modules, characterized by a loose, 34-amino acid consensus motif that is found in varying numbers of tandem repeats.[35] The N terminal TPR domain of Hop (TPR1) is required for Hsp70 binding[36] and a central TPR motif-containing region (TPR2A, Fig. 1A,B) is essential for Hsp90 binding.[37,38] The TPR-acceptor site on both Hsp70 and Hsp90 is comprised of an EEVD motif on the C-terminus.[38,39] It is also possible that there are networks of interactions between Hop and the chaperones Hsp70 and Hsp90, apart from those mediated by the TPR domains, which allow for its functionality as a scaffolding protein.[38]

Hop possesses insignificant chaperoning capabilities[40] and despite an increase in mRNA levels of mHop in mouse cells there is no change in the steady state levels of this protein following heat shock.[41] A similar occurrence has been described for human Hop (hHop) upon viral transformation.[29] Hop appears to be regulated between a monomeric and dimeric state, interacting with the dimeric Hsp90 as a dimer while associating with Hsp70 as a monomer.[42]

Initial studies of Hop focused on its role as an adaptor between Hsp70 and Hsp90 (Fig. 2) and the functioning of this multi-chaperone complex in steroid receptor (SR) regulation. SRs comprise of soluble intracellular proteins, which shuttle between the cytosol and the nucleus. They exist in an inactive or nontransformed state in the absence of their particular steroid hormone. Diffusion of the appropriate hormone into the cell transforms the receptor into an active transcription factor, which is capable of activating or repressing the expression of the steroid response genes.[43] The assembly of the progesterone receptor (PR) and the glucocorticoid receptor (GR) requires the participation of Hsp70, which brings the substrate protein into contact with Hsp90 via the scaffolding function of Hop. In the initial stage of the models proposed by both the Smith and Toft groups[42,44-46] Hsp40 binds to free SR, and facilitates the binding of the SR to Hsp70 through modulation of the ATPase cycle of Hsp70. Hip stabilizes this complex formation. Hsp70 then interacts with Hop, which is already in complex with Hsp90, and in this way, allows for the SR to come into contact with Hsp90. Hop is thus able to act as a "bridge" between the two major Hsps. This complex is generally referred to as the "intermediate complex".[46] Recent reports demonstrate the need for both TPR1 and TPR2 domains of yHop to be present on the same polypeptide, in order to maintain regulation of steroid receptor activation by Hsp70 and Hsp90.[47]

The intermediate step is followed by the release of Hop, Hsp70 and its co-chaperones Hsp40 and Hip and the formation of a mature complex that is stabilized by the presence of p23 and one of its TPR-containing immunophilins (immunophilins are a group of proteins which bind to immunosuppressive ligands; see Chapter by Cox and Smith). The result is a high affinity hormone binding conformation of the receptor. Hormone binding to the receptors releases them from Hsp90, and in the absence of bound hormone, dissociated receptor subunits reassociate with Hsp70 and proceed through the cycle again.[1,48]

Hop Modulates the Activities of Hsp70 and Hsp90

Some studies have suggested that Hop may change its conformation during the assembly of the Hsp70-Hop-Hsp90 chaperone heterocomplex, due to the fact that the affinity and

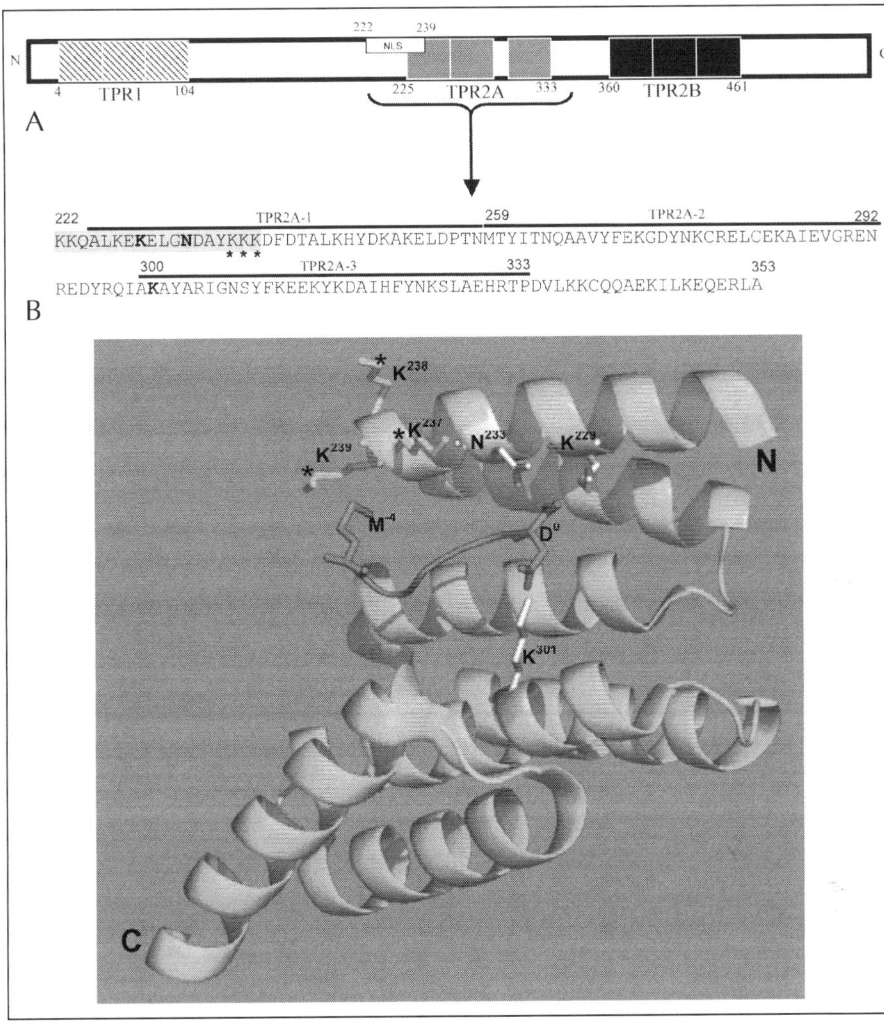

Figure 1. The Hop TPR2A domain overlaps with a putative NLS. A) Schematic diagram showing the TPR domain organization of Hop. The three TPR motifs which form part of the TPR1 domain are denoted by boxes with diagonal patterns, while those that form TPR2A and TPR2B domains are depicted as grey and black boxes, respectively. A solid white rectangle denotes the proposed NLS domain, which overlaps with TPR2A. TPR2A has been further enlarged diagrammatically, to show the amino acid sequence of this domain and a flanking C-terminal helix. TPR2A comprises of three TPR motifs (residues are shown by a thick black overline) denoted as TPR2A-1, TPR2A-2 and TPR2A-3.[80] The residues against a gray background are those of the proposed bipartite NLS[70,71] and which overlap with TPR2A-1 motif. The major arm of the bipartite NLS is denoted by an asterisk under each residue. The residues shown in bold, K229, N233 and K301 have been shown to be important for Hsp90 binding.[38] B) Ribbon representation of the structure of the TPR2A domain of hHop. The TPR2A domain and a flanking C-terminal helix (green) are shown interacting with the C-terminal MEEVD peptide (red) of Hsp90 (Protein Database code: 1ELR).[80] Residues shown in gold are those that are important for interaction with Hsp90 (refer to A). Residues denoted with an asterisk correspond to those that form the major arm of the bipartite NLS, which overlaps with the TPR2A domain. The figure was generated using Pymol Molecular Graphics Software (http://pymol.sourceforge.net).[81] A color version of this figure is available online at www.eurekah.com.

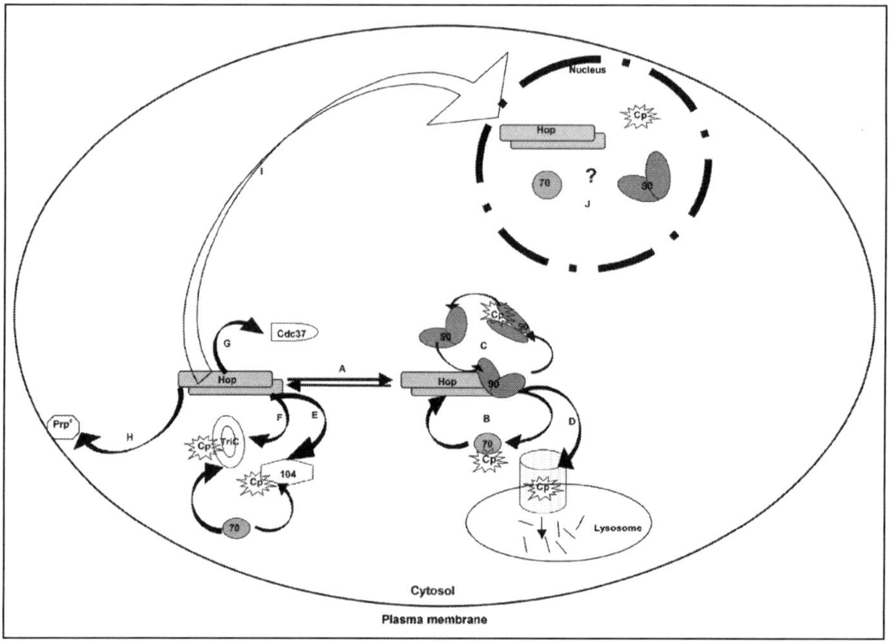

Figure 2. A summary of known cellular functions of Hop. Hop appears to exist either in the bound form to Hsp90, or as free Hop (A). B) reflects a simplistic view of its function as a scaffolding protein in bringing Hsp70 with the client protein into contact with Hsp90. Once the client protein is transferred onto Hsp90, Hop and Hsp70 dissociate, and the refolded client protein is then released, freeing Hsp90 (C). The Hsp70-Hop-Hsp90 chaperone complex is also involved in chaperone-mediated autophagy (D). Free Hop in yeast, interacts directly with Hsp104 (E), which is a stress tolerance factor that acts in concert with Hsp40 and Hsp70 to reactivate denatured proteins. Hop also interacts directly with TRiC (F), which is otherwise involved together with Hsp70, in the refolding of certain specific client proteins. Yeast Hop interacting with Cdc37 (G), a molecular chaperone and co-chaperone of Hsp90, appears to be important for Cdc37 to enter the chaperone dependent-folding pathway. Mouse Hop binds to Prp^C both in vitro and in vivo, and this interaction was found to transduce neuroprotective signals (H). Hop has been speculated to shuttle bi-directionally between the nucleus and cytoplasm (I). It is now known that both Hsp70 and Hsp90 also move into the nucleus during specific conditions. The nuclear role of Hop within its co-chaperoning context remains to be elucidated (J). Although Hop is generally depicted as a dimer, it must be noted that the stoichiometry of Hop in some of these interactions has not been determined. Heat shock proteins are identified by their molecular mass stated as numbers. CP stands for client protein. A color version of this figure is available online at www.eurekah.com.

stoichiometry of Hsp70-Hop binding is dramatically affected by binding of Hsp90. Although Hsp70 binds to Hop with a relatively lower affinity than Hsp90, this affinity is increased five fold in the presence of Hsp90.[42] Hsp90 may thus be altering the conformation of Hop to one that better accommodates interactions of Hop with Hsp70. Another possibility is that Hop binding to Hsp90 may open up a new conformation of Hsp90 that provides contact sites for Hsp70 binding.[42,49]

Despite findings that show that hHop and yHop are capable of functionally complementing for each other,[45,50] the involvement of Hop within the chaperone machineries of mammalian and yeast cells have shown some differences. Practically all of the yHop protein exists in a complex with Hsp90.[42,51] The basic elements of the Hsp90 chaperone complex in yeast are similar to that of vertebrates;[51] however it has been suggested that yHop is not essential for mediating associations between Hsp70, Hsp90 and target proteins in yeast.[52] This is in contrast to the mammalian system in which it has been shown that Hop is necessary for efficient assembly of steroid receptor-Hsp90 complexes in vitro[27,28,38] and that Hop is essential in integrating Hsp70-Hsp90 interactions.[45,47]

Morishima et al[53] reported that Hop, rather than being essential for GR folding by the Hsp90 based chaperone system, enhances the rate of this phenomenon. Furthermore, the same authors have shown that the protein levels of Hop were increased in geldanamycin-blocked GR complexes of Hsp90 compared to GR-Hsp90 complexes in the absence of geldanamycin. This may be due to the fact that geldanamycin-inhibited Hsp90 is in an ADP-bound conformation, which has higher affinity for Hop than the ATP-conformation.[53,54] In a Hop-depleted system, GR that was incubated with Hsp90 inhibitor geldanamycin, displayed little or no association with Hsp90, whereas the same system showed a more stable GR-Hsp90 association in the presence of Hop.[53] The same authors reported that the effect of Hop on GR-Hsp90 interactions in the presence of geldanamycin could either be due to the stable retention of Hsp90 by Hop, or due to some kind of influence that Hop may be exerting on the geldanamycin-Hsp90 conformation such that its affinity for GR is increased. Exclusion of Hop results in the reduced activity, but not accumulation, of two structurally and functionally unrelated Hsp90 client proteins, the steroid receptor GR and the oncogenic tyrosine kinase v-Src. The exclusion of Hop did not, however, have an effect on the activity of c-Src which is a protein closely related to v-Src but less dependent on Hsp90. This suggests that Hop is an important factor in promoting the maturation of Hsp90 client proteins.[52]

In the human system, hHop has no effect on Hsp70's ATPase activity, alone or in combination with Hsp90.[37] hHop does not affect Hsp90's ATPase cycle in the human system either, although it is capable of inhibiting client protein-stimulated ATPase activity of Hsp90.[55] yHop however, in direct contrast, stimulates ATP hydrolysis of Hsp70, enhancing its ATPase activity by a factor of 200 and is a noncompetitive inhibitor of Hsp90's ATPase activity.[56,57] Studies conducted by Wegele et al [56] showed that yHop is capable of accelerating ATP hydrolysis of Hsp70 to a greater extent than any other stimulation factor including yeast Hsp40. Even if yHop was added to a preformed Hsp70-Hsp40 complex, it was still able to activate the ATPase activity of Hsp70 and moreover, yeast Hsp40 was unable to replace yHop in a preformed Hsp70-Hop complex. Binding of yHop to Hsp90 and Hsp70 allowed activation of Hsp70 ATPase and inhibition of Hsp90 ATPase at the same time.[56]

The ATPase inhibition of Hsp90 by yHop is achieved by restricting N-terminal dimerization, which is a necessary conformational change in Hsp90 for ATP hydrolysis. This was confirmed by studies demonstrating a Hop binding site in the N-terminal region of Hsp90, in addition to the already characterized C-terminal peptide region that interacts with the TPR2A domain of Hop.[57] This Hop-mediated suppression of ATP turnover by Hsp90, is the motivation underpinning the postulation that Hop is involved in preparing Hsp90 for fresh "loading" of substrate protein.[24,58]

The Hsp70-Hop-Hsp90 multi-chaperone machinery is also involved in a process called "chaperone-mediated autophagy", one that targets cytosolic proteins to the lysosomes for degradation in response to stress conditions such as prolonged starvation or serum withdrawal (Fig. 2).[59] Protein substrates have to become unfolded in order to be transported into the lysosomal lumen. Hop is speculated to be part of the strategy employed to stabilize the lysosomal Hsp70-substrate complex on the lysosomal surface in such a way that it allows for the complete unfolding of the substrate protein before import into the lysosome.[59]

A recent report by Song and Masison [47] clearly demonstrates impairment of an Hsp70-dependent chaperone pathway upon deletion of the TPR1 domain in yHop, as well as an impaired Hsp90-dependent chaperone pathway upon deletion of the TPR2 domain of yHop. Deletion of TPR1 did not affect Hsp90-dependent client protein activity, and deletion of TPR2 also had no adverse effect on Hsp70-dependent client protein activity. These deletions, however, impaired client protein folding pathways that involved both Hsp70 and Hsp90.[47] This strongly indicates that Hop regulates Hsp70 and Hsp90 chaperone pathways independently, as well as concurrently. These findings strengthen the potential role of Hop as an active modulator of the functions of both Hsp70 and Hsp90, in addition to its passive role of serving as an "adaptor" between these two chaperone machineries.

Hop Interactions Go Beyond Hsp70 and Hsp90

Hsp70-Hop-Hsp90 interactions are the most well characterized TPR-mediated interactions of Hop, however protein-protein contacts through this domain are not exclusive to the Hsp70-Hsp90 multi-chaperone complex. Studies done with *Saccharomyces cerevisiae*, show that in the presence of nonfermentable carbon sources like ethanol and glycerol, yHop (in addition to other Hsp90 co-chaperones, Cpr7 and Cns1) interacts with Hsp104 through its N-terminal TPR1 domain (Fig. 2) and this interaction has been shown to be independent of Hsp90.[19] Hsp104 is a stress tolerance factor, which acts in concert with Hsp40 and Hsp70 to reactivate denatured proteins.[60]

yHop directly interacts with Cdc37 (Fig. 2), a co-chaperone of Hsp90, and this interaction is speculated to occur via both TPR1 and TPR2 domains of Hop but possibly not on the same binding sites as those involved in binding of Hop to both Hsp70 and Hsp90.[61] Recent work by Harst et al[62] confirmed this interaction with the mammalian homologue of Cdc37, p50, and suggested the presence of a complex that comprises of Hsp90, yHop and p50, in which one of the Hsp90 cofactors acted as the central component. yHop is not essential for growth of yeast cells at 30°C but growth impairment occurs at higher and lower temperatures or in the presence of minimal media.[26,27] Interestingly however, the combination of Cdc37 and yHop mutations is synthetically lethal to yeast under normal conditions, implying that their interaction may contribute to the vital functioning of yeast.[61] On the basis of Cdc37 being a molecular chaperone (see Chapter by Caplan), as well as the findings that prevention of aggregation of polypeptides and folding of protein kinases require the presence of Cdc37 as a co-chaperone to Hsp90, Lee et al[63] have speculated that yHop interaction with Cdc37 may be crucial for Cdc37 to enter the chaperone dependent-folding pathway.

Hop has also shown a direct interaction with the eukaryotic chaperonin-containing TCP1 (CCT), also known as the TCP-1 ring complex (TriC), which seems to be involved in the proper folding of actins and tubulins.[64] This is diagrammatically represented in Figure 2. While Hop showed no effect on the ATPase activity of CCT, it significantly stimulated nucleotide exchange, thereby interfering with substrate-associative capabilities of CCT. CCT cooperates with Hsp70 in refolding of luciferase in vitro, and this phenomenon is proposed to occur in vivo for certain substrates after translation or after stress-induced damage. The interaction of Hop with CCT was mediated through its C-terminal domain, in contrast to its interaction with Hsp70, which is mainly through its N-terminal domain, and this is consistent with the observation that the presence of Hsp70 did not affect Hop-CCT interactions.[64]

Hop contains, in addition to its TPR domains, two smaller domains with characteristic DP repeat motifs comprising four amino acid residues, reflected in an arrangement that corresponds to TPR1-DP1-TPR2A-TPR2B-DP2.[65,66] Recent comparison studies on hHop, yHop and *Drosophila melanogaster* Hop which lacks DP1 (dHop), showed that dHop cannot support GR function in yeast, although it can still bind to both Hsp70 and Hsp90, and can complement for, and thus rescue, growth defects in yeast which lack yHop.[66] Disruption of DP2 abrogates Hsp70 binding (implying an interaction between DP2 and TPR1). The substitution of DP2 of hHop by DP2 from dHop does not affect Hsp70 binding although it fails to support GR activity. A substitution of DP2 from dHop with DP2 from hHop regains the ability to enhance GR activity. It is possible therefore, that the DP2 domains may be responsible for an additional function of Hop to enhance GR activity, besides its Hsp70 binding capabilities. Carrigan et al[66] have thus proposed a novel role for Hop in GR maturation in vivo, which is independent of Hsp70/Hsp90 binding and showed using chimeric studies, that DP2 is critical for this "new" role of Hop.

Another novel role of Hop, in neuroprotection, was proposed by Martins' group.[67] Recombinant mHop was found to bind to Prp^c (a protein whose expression is crucial to the propagation of neurological disease, in particular, Prion disease) both in vitro and in vivo. The interaction of Prp^c with mHop was found to transduce neuroprotective signals (Fig. 2). A number of other molecules have shown in vitro association with Prp^c such as Hsp60, BiP, Bcl-2 and a 37/67 kDa laminin receptor,[68] but physiological relevance in the form of neuroprotection has only been attributed to the Prp^c – laminin complex. The laminin receptor-binding site on the Prp^c molecule maps to a region of amino acids that is significantly distinct from that of the mHop-binding domain. mHop may therefore participate within a Prp^c – laminin complex wherein association of Prp^c with both molecules may supply an additive effect.[67]

Subcellular Localization of Hop Affects Its Activities

The subcellular localization of Hop has definite implications on its various functions, both within, as well as outside the context of the Hsp70/Hsp90 chaperone heterocomplex. hHop has been found in the Golgi apparatus and small vesicles in normal cells, and nucleolar localization of hHop has been described in SV40-transformed cells.[29] In contrast, mHop has been described as being predominantly cytoplasmic with a small percentage of it being identified in the nucleus.[41] Nuclear Hop has been shown to be a crucial component of the OCA-S complex, involved in the regulation of S-phase dependent Histone H2B transcription.[69] An investigation into mHop's subcellular localization by Blatch's group described a predominantly nuclear accumulation of mHop under conditions of G1/S arrest or leptomycin B treatment in mouse fibroblast cells, leading to the proposal that there is a constant shuttling of the protein between nucleus and cytosol, with the export of mHop from the nucleus occurring at a faster rate than its import.[70] A proposed nuclear localization signal (NLS) in mHop, when fused to EGFP (enhanced green fluorescent protein), resulted in the localization of EGFP within the nucleus, suggesting that this NLS was functional in mHop.[70]

Interestingly, the proposed NLS overlaps with the TPR2A domain, which modulates interactions between Hop and Hsp90. Figure 1A shows a partial amino acid sequence of Hop, demonstrating the overlap of the proposed NLS with the TPR2A domain. The NLS, by virtue of its proximity relative to the TPR2A domain, may therefore contain residues involved in Hsp90 binding. A three dimensional representation of the crystal structure of TPR2A interacting with the pentapeptide MEEVD peptide of Hsp90 is presented in Figure 1B, displaying the proximity of the NLS relative to the sites on TPR2A involved in binding to Hsp90. Interactions of Hop with Hsp90, may therefore be involved in the mechanism of the nuclear localization of Hop. A possibility is that Hop binds a nuclear import factor like importin α and Hsp90 alternately and each interaction mediates nuclear import or cytosolic retention, respectively. Binding of Hsp90 to the TPR2A domain may mask the NLS preventing interactions with importin α and thus retaining Hop within the cytosol.

Acidic isoforms of Hop were elevated after viral transformation[29] and heat shock,[41] suggesting that this protein is phosphorylated during stress. There is evidence for the in vitro phosphorylation of mHop by casein kinase II (CKII; S189) and cdc2 kinase (T198) at sites located upstream of a putative NLS.[71] Furthermore, there is evidence that phosphorylation of Hop at these sites regulates its localization.[70] The postulation is that phosphorylation of mHop by CKII or cdc2 kinase promotes the nuclear import or cytosolic retention of mHop, respectively. It is therefore possible that phosphorylation of mHop at either or both of these sites may affect its interactions with Hsp90, thereby regulating the assembly of the Hsp70-Hsp90 chaperone heterocomplex.[72]

In light of Hop's subcellular localization and its interaction with Prp^c, it is interesting to note that wild-type Prp^c is not detected in the cytoplasm and is localized predominantly on cell surfaces or synaptosomal fractions.[67,73] Martins et al[74] reported the presence of a small fraction of the Prp^c ligand, which was unknown at the time, at the cell surface. Using membrane preparations from mouse brain, the same authors have shown through immunoprecipitation that at least part of mHop, which is postulated to be that "unknown" Prp^c-ligand, is localized at the cell surface, where it interacts with Prp^c.[67] It has been proposed that mHop is taken into the cell membrane as part of a protein complex, or secreted by a pathway that is distinct from the classical route through the ER and Golgi apparatus[75] due to the absence of a transmembrane domain or signal peptide for membrane transport within the amino acid sequence of mHop.[41,67]

Conclusion

Although the function of Hop has conventionally been restricted to that of Hsp70/Hsp90 organizing protein, it is apparent now that even within this system Hop has an active role to play as a modulator of their chaperone activities and protein folding pathways (Fig. 2). The presence of its various isoforms, the strong evidence for post-translational phosphorylation, a varied subcellular localization pattern and the possibility that its localization may be linked to post-translational modifications, strongly suggests complex roles for Hop in different systems and under different cellular conditions.

Members of the Hsp70 family migrate to the nucleus particularly during heat shock, where they are involved in stress-related cytoprotection. Hsp70 is capable of translocating nuclear proteins into the nucleolus during stress, possibly to prevent the random aggregation of thermolabile proteins within the nucleus and thereby preventing damage to other nuclear components.[76] Hsp70 is also known to regulate the activity of certain nuclear DNA-binding transcription factors.[77] Although Hsp90 is predominantly cytosolic, it is also known to translocate into the nucleus and associate with nuclear membranes, under conditions of stress,[15,78,79] thereby maintaining the integrity of the nuclear envelope and possibly other nuclear structures during heat shock.[79] The presence of Hop in the nucleus under prescribed conditions, and the possibility that it may be translocating to the nucleus via a functional NLS, is particularly intriguing and poses a number of questions as to what its functions are within the nucleus. Reports have already been published regarding the involvement of nuclear Hop in complex with Hsp70, in cell cycle-regulated transcription of histone H2B.[69] Considering the potential variety of roles and complexes of Hop in the cytosol, it is likely that Hop may also possess a varied role within the nucleus and may be interacting within a number of other nuclear complexes. However, these proposed new roles of Hop remain to be fully elucidated.

Interactions of Hop with structurally and functionally unrelated proteins makes it increasingly difficult to define Hop as merely a Hsp70/Hsp90 adaptor or co-chaperone, and presents the multifaceted nature of its biological functions (Fig. 2). Questions regarding the actual mechanism(s) by which Hop is able to distinguish between its different interactions as well as its subcellular localization, need to be answered in order to gain further insight into its global function within the biological system.

Acknowledgements

We gratefully acknowledge funding for this research from the National Research Foundation (NRF) Unlocking the Future Focus Area and NRF-Hungary Collaborative Programme (NRF, South Africa, GUN Nos: 2067467 and 2053542). SD was awarded a NRF grant-holder doctoral bursary, and a Rhodes University doctoral scholarship. The authors would like to acknowledge Dr. Sophie Jackson (University of Cambridge, UK) for her valuable comments on this manuscript.

References

1. Smith DF, Whitesell L, Katsanis E. Molecular chaperones: Biology and prospects for pharmacological intervention. Pharmacol Rev 1998; 50:493-513.
2. Morimoto RI. Regulation of the heat shock transcriptional response: Cross talk between a family of heat shock factors, molecular chaperones, and negative regulators. Genes Dev 1998; 12:3788-3796.
3. Frydman J. Folding of newly translated proteins in vivo: The role of molecular chaperones. Annu Rev Biochem 2001; 70:603-647.

4. Macario AJL, de Macario EC. The archaeal molecular chaperone machine: Peculiarities and paradoxes. Genetics 1999; 152:1277-1283.
5. Bukau B, Deuerling E, Pfund C et al. Getting newly synthesized proteins into shape. Cell 2000; 101:119-122.
6. Ryan MT, Pfanner N. Hsp70 proteins in protein translocation. Adv Protein Chem 2001; 59:223-242.
7. Hartl FU, Hayer-Hartl M. Molecular chaperones in the cytosol: From nascent chain to folded protein. Science 2002; 295:1852-1858.
8. Fink AL. Chaperone-mediated protein folding. Physiol Rev 1999; 79:425-449.
9. Mayer MP, Schröder H, Rüdiger S et al. Multistep mechanism of substrate binding determines chaperone activity of Hsp70. Nat Struct Biol 2000; 7:586-593.
10. Liberek K, Marszalek J, Ang D et al. Escherichia coli DnaJ and GrpE heat shock proteins jointly stimulate ATPase activity of DnaK. Proc Natl Acad Sci USA 1991; 88:2874-2878.
11. Sondermann H, Scheufler C, Schneider C et al. Structure of a Bag/Hsc70 complex: Convergent functional evolution of Hsp70 nucleotide exchange factors. Science 2001; 291:1553-1557.
12. Höhfeld J, Minami Y, Hartl FU. Hip, a novel co-chaperone involved in the eukaryotic Hsc70/Hsp40 reaction cycle. Cell 1995; 83:589-598.
13. Csermely P, Schnaider T, Söti C et al. The 90-kDa molecular chaperone family: Structure, function, and clinical applications. A comprehensive review. Pharmacol Ther 1998; 79:129-168.
14. Louvion JF, Abbas-Terki T, Picard D. Hsp90 is required for pheromone signaling in yeast. Mol Biol Cell 1998; 9:3071-3083.
15. Pratt WB, Toft DO. Steroid receptor interactions with heat shock protein and immunophilin chaperones. Endocr Rev 1997; 18:306-360.
16. Pratt WB. The role of hsp90-based chaperone system in signal transduction by nuclear receptors and receptors signaling via MAP kinase. Annu Rev Pharmacol Toxicol 1997; 37:297-326.
17. Panaretou B, Prodromou C, Roe SM et al. ATP binding and hydrolysis are essential to the function of the Hsp90 molecular chaperone in vivo. EMBO J 1998; 17:4829-4836.
18. Zhao R, Davey M, Hsu YC et al. Navigating the chaperone network: An integrative map of physical and genetic interactions mediated by the Hsp90 chaperone. Cell 2005; 120:715-727.
19. Abbas-Terki T, Donzé O, Briand PA et al. Hsp104 interacts with Hsp90 co-chaperones in respiring yeast. Mol Cell Biol 2001; 21:7569-7575.
20. Csermely P, Kahn CR. The 90-kDa heat shock protein (hsp-90) possesses an ATP binding site and autophosphorylating activity. J Biol Chem 1991; 266:4943-4950.
21. Young JC, Hartl FU. Polypeptide release by Hsp90 involves ATP hydrolysis and is enhanced by the co-chaperone p23. EMBO J 2000; 19:5930-5940.
22. Csermely P, Kájtar J, Hollósi M et al. ATP induces a conformational change of the 90-kDa heat shock protein (hsp90). J Biol Chem 1993; 268:1901-1907.
23. Söti C, Rácz A, Csermely P. A nucleotide-dependent molecular switch controls ATP binding at the C-terminal domain of Hsp90. J Biol Chem 2002; 277:7066-7075.
24. Siligardi G, Panaretou B, Meyer P et al. Regulation of Hsp90 ATPase activity by the co-chaperone Cdc37p/p50^{cdc37}. J Biol Chem 2002; 277:20151-20159.
25. Panaretou B, Siligardi G, Meyer P et al. Activation of the ATPase activity of Hsp90 by the stress-regulated co-chaperone Aha1. Mol Cell 2002; 10:1307-1318.
26. Nicolet CM, Craig EA. Isolation and characterization of STI1, a stress-inducible gene from Saccharomyces cerevisiae. Mol Cell Biol 1989; 9:3638-3646.
27. Smith DF, Sullivan WP, Marion TN et al. Identification of a 60-kilodalton stress-related protein, p60, which interacts with hsp90 and hsp70. Mol Cell Biol 1993; 13:869-876.
28. Dittmar KD, Hutchison KA, Owens-Grillo JK et al. Reconstitution of the steroid receptor.hsp90 hetereocomplex assembly system of rabbit reticulocyte lysate. J Biol Chem 1996; 271:12833-12839.
29. Honoré B, Leffers H, Madsen P et al. Molecular cloning and expression of a transformation-sensitive human protein containing the TPR motif and sharing identity to the stress-inducible yeast protein STI1. J Biol Chem 1992; 267:8485-8491.
30. Blatch GL, Lässle M, Zetter BR et al. Isolation of a mouse cDNA encoding mSTI1, a stress-inducible protein containing the TPR motif. Gene 1997; 194:277-282.
31. Demand J, Lüders J, Höhfeld J. The carboxyl-terminal domain of Hsc70 provides binding sites for a distinct set of chaperone cofactors. Mol Cell Biol 1998; 18:2023-2028.
32. Adams MD, Celniker SE, Holt RA et al. The genome sequence of Drosophila melanogaster. Science 2000; 287:2185-2195.
33. Zhang Z, Quick MK, Kanelakis KC et al. Characterization of a plant homolog of Hop, a co-chaperone of Hsp90. Plant Physiol 2003; 131:525-535.

34. Webb JR, Campos-Neto A, Skeiky YAW et al. Molecular characterization of the heat-inducible LmSTI1 protein of Leishmania major. Mol Biochem Parasitol 1997; 89:179-193.
35. Blatch GL, Lässle M. The tetratricopeptide repeat: A structural motif mediating protein-protein interactions. BioEssays 1999; 21:932-939.
36. van der Spuy J, Kana BD, Dirr HW et al. Heat shock cognate protein 70 chaperone-binding site in the co-chaperone murine stress-inducible protein 1 maps to within three consecutive tetratricopeptide repeat motifs. Biochem J 2000; 345:645-651.
37. Chen S, Prapapanich V, Rimerman RA et al. Interactions of p60, a mediator of progesterone receptor assembly, with heat shock proteins Hsp90 and Hsp70. Mol Endocrinol 1996; 10:682-693.
38. Odunuga OO, Hornby JA, Bies C et al. Tetratricopeptide repeat motif-mediated Hsc70-mSTI1 interaction: Molecular characterization of the critical contacts for successful binding and specificity. J Biol Chem 2003; 278:6896-6904.
39. Brinker A, Scheufler C, von der Mülbe F et al. Ligand discrimination by TPR domains: Relevance and selectivity of EEVD-recognition in Hsp70.Hop.Hsp90 complexes. J Biol Chem 2002; 277:19265-19275.
40. Schumacher RJ, Hurst R, Sullivan WP et al. ATP-dependent chaperoning activity of reticulocyte lysate. J Biol Chem 1994; 269:9493-9499.
41. Lässle M, Blatch GL, Kundra V et al. Stress-inducible murine protein mSTI1: Characterization of binding domains for heat shock proteins and in vitro phosphorylation by different kinases. J Biol Chem 1997; 272:1876-1884.
42. Hernández MP, Sullivan WP, Toft DO. The assembly and intermolecular properties of the hsp70-Hop-hsp90 molecular chaperone complex. J Biol Chem 2002; 277:38294-38304.
43. Scherrer LC, Dalman FC, Massa E et al. Structural and functional reconstitution of the glucocorticoid receptor-Hsp90 complex. J Biol Chem 1990; 265:21397-21400.
44. Smith DF, Whitesell L, Nair SC et al. Progesterone receptor structure and function altered by geldanamycin, an hsp90-binding agent. Mol Cell Biol 1995; 15:6804-6812.
45. Chen S, Smith DF. Hop as an adaptor in the heat shock protein 70 (Hsp70) and Hsp90 chaperone machinery. J Biol Chem 1998; 273:35194-35200.
46. Smith DF, Toft DO. Steroid receptors and their associated proteins. Mol Endocrinol 1993; 7:4-11.
47. Song Y, Masison DC. Independent regulation of Hsp70 and Hsp90 chaperones by Hsp70/90 organizing protein Sti1 (Hop1). J Biol Chem 2005; 280:34178-34185.
48. Wegele H, Müller L, Buchner J. Hsp70 and Hsp90- a relay team for protein folding. Rev Physiol Biochem Pharmacol 2004; 151:1-44.
49. Murphy PJM, Kanelakis KC, Galigniana MD et al. Stoichiometry, abundance, and functional significance of the hsp90/hsp70-based multiprotein chaperone machinery in reticulocyte lysate. J Biol Chem 2001; 276:30092-30098.
50. Carrigan PE, Nelson GM, Roberts PJ et al. Multiple domains of the co-chaperone Hop are important for Hsp70 binding. J Biol Chem 2004; 279:16185-16193.
51. Chang HCJ, Lindquist S. Conservation of Hsp90 macromolecular complexes in Saccharomyces cerevisiae. J Biol Chem 1994; 269:24983-24988.
52. Chang HCJ, Nathan DF, Lindquist S. In vivo analysis of the Hsp90 co-chaperone Sti1 (p60). Mol Cell Biol 1997; 17:318-325.
53. Morishima Y, Kanelakis KC, Silverstein AM et al. The Hsp organizer protein Hop enhances the rate of but is not essential for glucocorticoid receptor folding by the multiprotein Hsp90-based chaperone system. J Biol Chem 2000; 275:6894-6900.
54. Sullivan W, Stensgard B, Caucutt G et al. Nucleotides and two functional states of hsp90. J Biol Chem 1997; 272:8007-8012.
55. McLaughlin SH, Smith HW, Jackson SE. Stimulation of the weak ATPase activity of human Hsp90 by a client protein. J Mol Biol 2002; 315:787-798.
56. Wegele H, Haslbeck M, Reinstein J et al. Sti1 is a novel activator of the Ssa proteins. J Biol Chem 2003; 278:25970-25976.
57. Richter K, Muschler P, Hainzl O et al. Sti1 is a noncompetitive inhibitor of the Hsp90 ATPase: Binding prevents the N-terminal dimerization reaction during the ATPase cycle. J Biol Chem 2003; 278:10328-10333.
58. Prodromou C, Siligardi G, O'Brien R et al. Regulation of Hsp90 ATPase activity by tetratricopeptide repeat (TPR)-domain co-chaperones. EMBO J 1999; 18:754-762.
59. Agarraberes FA, Dice JF. A molecular chaperone complex at the lysosomal membrane is required for protein translocation. J Cell Sci 2001; 114:2491-2499.
60. Glover JR, Lindquist S. Hsp104, Hsp70, and Hsp40: A novel chaperone system that rescues previously aggregated proteins. Cell 1998; 94:73-82.

61. Abbas-Terki T, Briand PA, Donzé O et al. The Hsp90 co-chaperones Cdc37 and Sti1 interact physically and genetically. Biol Chem 2002; 383:1335-1342.
62. Harst A, Lin H, Obermann WMJ. Aha1 competes with Hop, p50 and p23 for binding to the molecular chaperone Hsp90 and contributes to kinase and hormone receptor activation. Biochem J 2005; 387:789-796.
63. Lee P, Shabbir A, Cardozo C et al. Sti1 and Cdc37 can stabilize Hsp90 in chaperone complexes with a protein kinase. Mol Biol Cell 2004; 15:1785-1792.
64. Gebauer M, Melki R, Gehring U. The chaperone cofactor Hop/p60 interacts with the cytosolic chaperonin-containing TCP-1 and affects its nucleotide exchange and protein folding activities. J Biol Chem 1998; 273:29475-29480.
65. Prapapanich V, Chen S, Smith DF. Mutation of Hip's carboxy-terminal region inhibits a transitional stage of progesterone receptor assembly. Mol Cell Biol 1998; 18:944-952.
66. Carrigan PE, Riggs DL, Chinkers M et al. Functional comparison of human and Drosophila Hop reveals novel role in steroid receptor maturation. J Biol Chem 2005; 280:8906-8911.
67. Zanata SM, Lopes MH, Mercadante AF et al. Stress-inducible protein 1 is a cell surface ligand for cellular prion that triggers neuroprotection. EMBO J 2002; 21:3307-3316.
68. Martins VR, Mercadante AF, Cabral ALB et al. Insights into the physiological function of cellular prion protein. Braz J Med Biol Res 2001; 34:585-595.
69. Zheng L, Roeder RG, Luo Y. S phase activation of the histone H2B promoter by OCA-S, a coactivator complex that contains GAPDH as a key component. Cell 2003; 114:255-266.
70. Longshaw VM, Chapple JP, Balda MS et al. Nuclear translocation of the Hsp70/Hsp90 organizing protein mSTI1 is regulated by cell cycle kinases. J Cell Sci 2004; 117:701-710.
71. Longshaw VM, Dirr HW, Blatch GL et al. The in vitro phosphorylation of the co-chaperone mSTI1 by cell cycle kinases substantiates a predicted casein kinase II-p34^{cdc2}-NLS (CcN) motif. Biol Chem 2000; 381:1133-1138.
72. Odunuga OO, Longshaw VM, Blatch GL. Hop: More than an Hsp70/Hsp90 adaptor protein. BioEssays 2004; 26:1058-1068.
73. Kretzschmar HA, Tings T, Madlung A et al. Function of Prp(C) as a copper-binding protein at the synapse. Arch Virol Suppl 2000; 16:239-249.
74. Martins VR, Graner E, Garcia-Abreu J et al. Complementary hydropathy identifies a cellular prion protein receptor. Nat Med 1997; 3:1376-1382.
75. Muesch A, Hartmann E, Rohde K et al. A novel pathway for secretory proteins? Trends Biochem Sci 1990; 15:86-88.
76. Nollen EAA, Salomons FA, Brunsting JF et al. Dynamic changes in the localization of thermally unfolded nuclear proteins associated with chaperone-dependent protection. Proc Natl Acad Sci USA 2001; 98:12038-12043.
77. Diehl JA, Yang W, Rimerman RA et al. Hsc70 regulates accumulation of cyclin D1 and cyclin D1-dependent protein kinase. Mol Cell Biol 2003; 23:1764-1774.
78. Biggiogera M, Tanguay RM, Marin R et al. Localization of heat shock proteins in mouse male germ cells: An immunoelectron microscopical study. Exp Cell Res 1996; 229:77-85.
79. Langer T, Rosmus S, Fasold H. Intracellular localization of the 90 kDa heat shock protein (HSP90α) determined by expression of a EGFP–HSP90α-fusion protein in unstressed and heat stressed 3T3 cells. Cell Biol Int 2003; 27:47-52.
80. Scheufler C, Brinker A, Bourenkov G et al. Structure of TPR domain–peptide complexes: Critical elements in the assembly of the Hsp70-Hsp90 multichaperone machine. Cell 2000; 101:199-210.
81. DeLano WL. The PyMOL molecular graphics system. South San Francisco, CA, USA: DeLano Scientific LLC, 2005, (http://www.pymol.org.).

Chapter 4

Do Hsp40s Act as Chaperones or Co-Chaperones?

Meredith F.N. Rosser and Douglas M. Cyr*

Abstract

The Hsp70 family plays an essential role in cellular protein metabolism by acting as a polypeptide binding and release factor that interacts with nonnative regions of proteins at different stages of their life cycles.[1] Hsp40 proteins not only act as co-chaperones to facilitate complex formation between Hsp70 and client proteins, but it has also been proposed that Hsp40s use an intrinsic chaperone activity to bind and deliver the nonnative substrates to Hsp70. Herein, we review genetic, biochemical and structural data that describes the mechanisms by which Type I and Type II Hsp40 proteins act to bind substrates. The manner by which the functions of the Type I and Type II proteins are specified is also discussed.

Introduction

The Hsp40 family of proteins has been shown to play a role in a multitude of cellular processes including protection from cellular stress, folding of nascent polypeptides, refolding of denatured or aggregated proteins, disassembly of protein complexes, inhibition of polyglutamine aggregation, protein degradation, and protein translocation across membranes.[2-9] There are over 100 different Hsp40 family members with 44 Hsp40 genes present in the human genome alone and 20 Hsp40s identified in the *Saccharomyces cerevisiae* genome.[9,10] These proteins were identified as co-chaperones that stimulate the ATPase activity of the Hsp70 family of proteins (Fig. 1), but as we will see in the following discussion, there is more to the Hsp40s than this one function alone. There are three classes of Hsp40 proteins which are divided on the basis of their domain structure (Fig. 2).[11] Type I Hsp40s are descendants of *E. coli* DnaJ and contain the J domain, followed by a glycine/phenylalanine-rich region (G/F-rich), a zinc finger like region (ZFLR), and a conserved C-terminal domain (CTD) which plays an important role in dimerization. The Type II Hsp40's are similar to the type I Hsp40s but instead of the zinc finger like region they contain an extended G/F-rich region or a glycine/methionine (GM)-rich region and a CTD1 domain. Type III Hsp40s contain the J-domain but none of the other conserved domains found in Type I or II Hsp40s. Instead, they often have specialized domains that localize them to certain areas of the cell and provide specificity in substrate binding.[12,13] The Hsp40s are conserved across species and are found in organisms from bacteria to humans, and there are a variety of Type I, Type II, and Type III Hsp40s found in different cellular organelles where they can play specialized roles.[9,14] In order to better understand the cellular processes that involve these proteins, we first need to understand the

*Corresponding Author: Douglas M. Cyr—Department of Cell and Developmental Biology School of Medicine, University of North Carolina at Chapel Hill, Chapel Hill, North Carolina, 27599, U.S.A. Email: dmctr@med.unc.edu

Networking of Chaperones by Co-Chaperones, edited by Gregory L. Blatch.

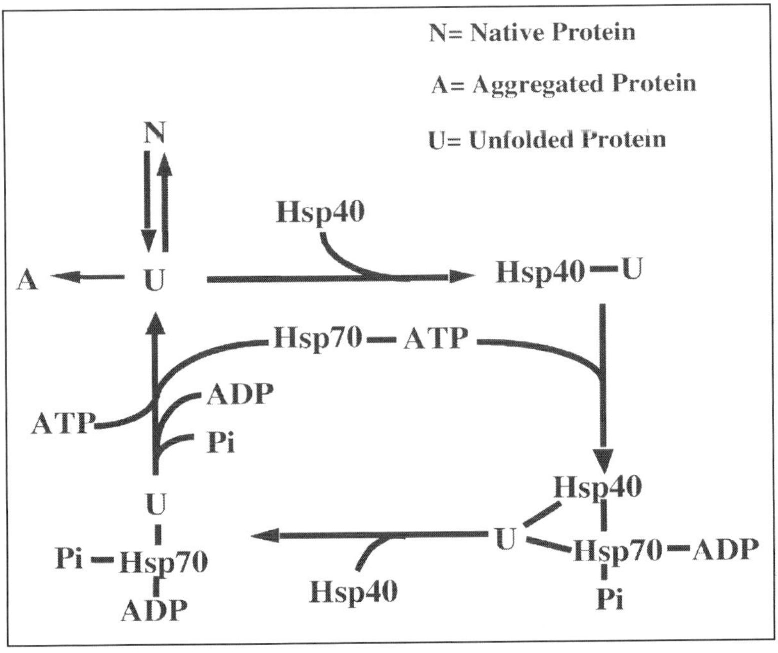

Figure 1. Model for regulation of Hsp70 folding cycle by Hsp40. Hsp70 has low substrate affinity in the ATP bound state but upon hydrolysis of ATP stable Hsp70-substrate complexes are formed. Hsp70-substrate complexes then disassociate upon regeneration of Hsp70-ATP. In this model, Hsp40 acts to (1) deliver substrates to Hsp70 and (2) stimulate the ATPase activity of Hsp70. This cycle is repeated numerous times until the substrate protein is able to reach a native state. Reprinted with permission from Fan CY et al, Cell Stress Chaperones 2003; 8(4):309-316.[14]

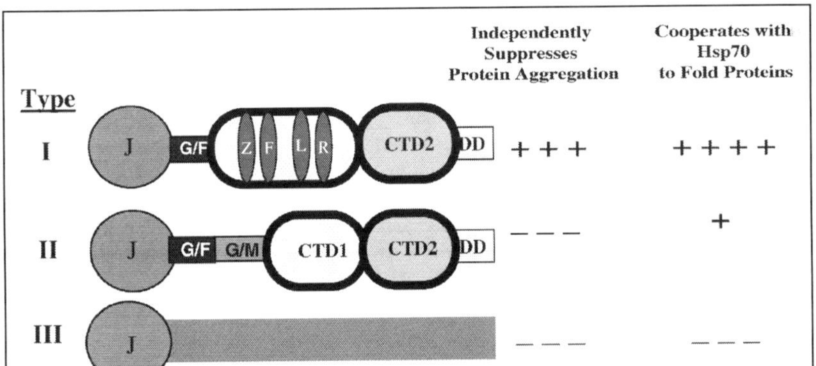

Figure 2. Domain structures of different Hsp40 subtypes. J: J-domain; G/F: glycine/phenylalanine rich region; ZFLR: zinc finger-like region; G/M: glycine/methionine rich region; CTD1: carboxyl-terminal domain 1; CTD2: carboxyl-terminal domain 2; DD: dimerization domain. The pluses and minuses represent whether or not members of each type of sub-family has been shown to possess the indicated activity. Reprinted wtih permission from Fan CY et al, Cell Stress Chaperones 2003; 8(4):309-316.[14] A color version of this figure is available online at www.eurekah.com.

mechanism by which these different types of Hsp40s function to bind substrates and to interact with and thereby regulate specific Hsp70 proteins. In the following sections, we will review the genetic, biochemical, cell biological, and structural data that have helped elucidate the unique mechanisms that different Hsp40s use to play such vital roles in cellular physiology.

Hsp70 Co-Chaperone Activity of Hsp40s

Similar to the Hsp40 family, the Hsp70 chaperones play a variety of essential physiological roles in the cell including assisting with protein folding and assembly, refolding of misfolded or aggregated proteins, degradation of misfolded proteins, and driving translocation of proteins across membranes.[8,15,16] In all of these processes, the affinity of Hsp70 for substrate is regulated by its nucleotide bound state. In the ATP bound form, Hsp70 has a low affinity for substrate proteins. However, upon hydrolysis of the ATP to ADP, Hsp70 undergoes a conformational change which increases its affinity for substrate proteins (Fig. 1). In order for Hsp70 to act as a chaperone and help substrate proteins fold, it goes through repeated cycles of ATP hydrolysis and nucleotide exchange thereby binding and releasing the substrate multiple times until it has reached the desired state.[8,15]

The Hsp70 proteins are assisted and regulated by several different co-chaperones. These co-chaperones have been shown to not only regulate different steps of the ATPase cycle of Hsp70 (Fig. 1), but they also have an individual specificity such that one co-chaperone may promote folding of a substrate while another may promote degradation. For example, the co-chaperones CHIP and Bag-1 are both thought to promote the degradation of Hsp70 bound substrates (see Chapter by Höhfeld et al).[17] On the other hand, the co-chaperones Hip and Hop are thought to promote the folding of Hsp70 substrates (see chapters by Höhfeld et al and Daniel et al).[17-19] Hsp40 proteins have been shown to play a role in both Hsp70 dependent folding and degradation pathways,[14,20-22] thereby suggesting a unique role in the regulation of Hsp70 activity. The Hsp40 proteins are classified as co-chaperones for Hsp70 due to the fact that they can use their various domain structures to (1) bind Hsp70 (2) help load the substrates on Hsp70 and (3) stimulate the ATPase activity of Hsp70. The action of Hsp40 to help load substrates on Hsp70 explains the ability of Hsp40 to participate in such a wide variety of cellular processes including steps in both folding and degradation pathways. The mechanism by which Hsp40s bind and interact with Hsp70s has been reviewed in great detail.[14,23,24] Therefore in the remainder of this chapter we will focus on the question of how Hsp40s bind substrates, and how this activity may specify any autonomous chaperone activity of the different Hsp40s.

Do Hsp40s Act as Chaperones?

It has been well established that Hsp40s can cooperate with Hsp70s in a variety of cellular processes. However, the exact manner by which the Hsp40s cooperate with the Hsp70s is not completely understood. One question about Hsp40 function is whether or not this family of proteins plays a role in cellular processes strictly by acting as a co-chaperone and regulating the affinity of Hsp70s for substrate proteins (Fig. 1), or if the Hsp40s use an intrinsic chaperone activity of their own to regulate this diverse multitude of cellular processes. If Hsp40s strictly act as co-chaperones, then they may participate in the loading of substrates on Hsp70s simply by stimulating the ATPase activity of the Hsp70s to convert the Hsp70 to a higher affinity conformation. However, if Hsp40s do have an intrinsic chaperone activity to bind nonnative substrates, then the Hsp40 can play a more active role in the Hsp70 dependent folding process by physically binding and delivering the substrates to Hsp70 as well as stimulating the ATPase activity. There is evidence that will be discussed below that Type I and Type II Hsp40s can act as independent chaperones, while Type III Hsp40s likely play more of a co-chaperone role. Some Type III Hsp40s do have individualized polypeptide binding domains, but they are generally considered to not bind denatured substrates and therefore are not considered to have any specific chaperone activity of their own.

The first observations of intrinsic chaperone activity of an Hsp40 came from studying the bacterial type I Hsp40, DnaJ.[25,26] DnaJ was first shown to have the ability to directly bind substrate in a study that showed the copurification of a 1:1 stoichiometric DnaJ-RepA complex from *E. coli*.[25] Another pivotal study of the DnaJ/DnaK complex showed that the DnaJ protein alone was able to suppress the aggregation of denatured rhodanese more effectively than the Hsp70, DnaK, alone.[26] These studies were the first to show that DnaJ could not only bind denatured substrates, but it could also prevent the aggregation of those denatured substrates, thereby categorizing the Hsp40s as chaperones in their own right.

Future studies bolstered the idea of intrinsic chaperone activity of Hsp40s when it was shown that DnaJ could also bind nascent polypeptides[27] and that DnaJ could bind denatured luciferase in an early step, followed by the formation of a ternary complex with DnaK.[28] In these experiments, the denatured luciferase was not efficiently refolded unless DnaJ bound first.[28] This chaperone ability is not limited to the DnaJ protein, and in fact numerous Type I Hsp40s from different organisms and different cellular locations retain the ability to both bind denatured substrates and suppress the aggregation of those substrates.[29-31] For example, Ydj1, a yeast type I Hsp40, was shown to suppress aggregation of denatured rhodanese and chemically or heat denatured luciferase,[31-33] and ERdj3, the human DnaJ homolog found in the ER, was shown to be able to directly bind multiple unfolded substrates irrespective of an interaction with BiP, the ER Hsp70.[30] Therefore, it is clear that the Type I chaperones can bind denatured substrates and prevent the aggregation of those substrates. Once the Type I Hsp40s bind substrate, it has been proposed that they then directly transfer that substrate to an Hsp70. In such a scenario, a ternary complex formed between the Hsp40, substrate, and Hsp70, would be an important intermediate step. In fact, DnaJ:polypeptide:DnaK ternary complexes have been isolated and the formation of such complexes appears to facilitate substrate transfer from Hsp40 to Hsp70.[34]

The story for the Type II Hsp40s has not been as clear. Studies with the yeast Sis1 protein have shown that Sis1 can bind chemically denatured luciferase and reduced α-lactalbumin and that this binding is dependent on specific residues within the C-terminal peptide binding domain.[35] This ability of Sis1 to recognize and bind nonnative polypeptides classifies Sis1 as a chaperone. However, Sis1 alone is not as effective a chaperone as the Type I Hsp40s because Sis 1 can not prevent the aggregation of thermally denatured luciferase nor does it hold the thermally denatured luciferase in a folding competent state.[33] However, Sis1 is able to hold chemically denatured luciferase in a folding competent state.[36] The human Hsp40, Hdj-1, also retains a weak ability to bind denatured luciferase,[37] and is also able to cooperate with Hsp70 to refold luciferase.[38] However, as seen with Sis1, the Hdj-1 alone is unable to hold thermally denatured luciferase in a folding competent state.[38] Other studies showed that Hdj-1 was unable to bind denatured β-galactosidase with a high enough affinity to gel shift the β-galactosidase into a native gel, but it could stimulate the ability of Hsp70 to refold the denatured β-galactosidase.[39] These studies would suggest that Type II Hsp40s do not bind denatured substrates with as high an affinity as the Type I Hsp40s. However, recent studies by the Van Bennett group have shown that Hdj-1 can bind the C-terminal domain of ankyrin-B with nanomolar affinity, and deletion of the Hdj-1 binding site results in loss of function of the ankyrin-B.[40]

The observed differences in the independent chaperone activity of the type I and type II Hsp40s may come from their ability to select different substrates. Perhaps the Type II Hsp40s recognize different amino acids combinations or conformations than the Type I Hsp40s such that they may bind one substrate with a lower affinity than the Type I Hsp40, but other substrates, such as the ankyrin-B, with a higher affinity. In fact, studies of the substrate specificity of Type I and Type II Hsp40s do show that there are subtle differences between the two types of chaperones. In order to determine what type of structure the Hsp40 chaperones recognize, studies first used the bacterial Hsp40, DnaJ, to determine whether it shows a preference for peptides consisting of either L- or D-amino acids.[41,42] The results showed that DnaJ could bind peptides consisting of both L- and D-amino acids thereby suggesting that Hsp40s rely more on side-chain interactions than backbone recognition to bind substrates. This is different than to the way Hsp70 chaperones have been shown to recognize substrates since substrate binding by Hsp70 involves both side-chain and backbone contacts.[43] It has been proposed that the ability of an Hsp40 to recognize amino acids side chains may enable it to scan substrates for hydrophobic surfaces and make the initial contacts with proteins that are subsequently targeted to Hsp70.[42] In order to determine if other Type I Hsp40s use a similar mechanism to bind substrates, selectivity was compared between DnaJ and Ydj1. DnaJ and Ydj1 both bind peptides enriched in the aromatic amino acids F, W and Y, the large hydrophobics I and L and the polar residue H. In addition Ydj1 and DnaJ appear to exclude the amino acids P and K from the peptides they select.[42,44]

Studies have also utilized peptide arrays to do a side by side comparison between Type I and Type II Hsp40s. Ydj1 and Sis1 were used to biopan a 7-mer phage peptide display library,[44] and the results showed that both Hsp40s selected peptides which were enriched in aromatic and bulky hydrophobic amino acids. However, the groups of peptides that were selected by Ydj1 and Sis1 exhibited slight differences in the enrichment of specific amino acids such that Ydj1 preferred peptides that had a hydrophobic stretch of 3-4 residues, but peptides selected by Sis1 did not contain a patch of hydrophobic residues.[44] Thus, while Type I and Type II Hsp40s can both bind hydrophobic residues, there are slight differences in their substrate specificity that may lend to their differences in chaperone ability.

Direct comparisons between the Type I and Type II Hsp40s show that not only are the Type I proteins more effective chaperones, but they also are more effective at cooperating with Hsp70 to refold denatured substrates. For example, Ydj1 can cooperate with Ssa1 to refold a higher percentage of denatured substrate than a Sis1: Ssa1 pair.[33] This same trend where the Type I Hsp40 is more effective than the Type II Hsp40 is also seen with the human Hdj-2 and Hdj-1 proteins.[2,37] The Type I protein, Hdj-2, binds denatured luciferase better than the type II protein, Hdj-1.[37] In one study, a direct comparison of Hdj-1 and Hdj-2 showed that under those experimental conditions, the Hdj-2:Hsp70 pair could effectively refold denatured luciferase, while the combination of Hsp70 and Hdj-1 could not.[2] However, other studies[45] have shown that Hdj-1 can cooperate with Hsp70 to refold denatured luciferase. So, there is some ambiguity in the data as to the ability of Hdj1.

Overall, the data do suggest that Type I Hsp40s are not only better independent chaperones, but they are also more efficient at cooperating with Hsp70 to refold substrates than the Type II Hsp40s.[2,33] Whether or not this is due to the intrinsic chaperone ability of the different types of Hsp40s, or due to different co-chaperone abilities is still not clear. The ability of Type I Hsp40s to protect nonnative substrates from aggregation may keep a higher percentage of substrate in a folding competent state, or it is also possible that the Type I and Type II Hsp40s do not transfer substrates to Hsp70 with the same efficiency. It is also possible that the above results could be explained by the idea that certain Hsp40s may functionally partner with only specific Hsp70s and not others.

Interestingly, while Type II Hsp40s do have diminished chaperone capacity in relation to the Type I Hsp40s, it is the yeast Type II Hsp40, Sis1 that is essential for viability,[10,46,51] while deletion of the yeast Type I Hsp40, Ydj1, only causes growth defects.[47] Studies have also shown that Ydj1 cannot substitute for Sis 1 function.[46] This lends to the obvious question of whether it is the chaperone or co-chaperone functions of the Hsp40s that are essential in vivo. To address this question, yeast were engineered to express Ydj1 or Sis1 mutants that do contain the J domain, which is responsible for stimulating Hsp70's ATPase activity, but do not contain the substrate binding domains.[48,49] The original studies showed that these mutants were completely able to sustain growth of the yeast at normal temperatures, thereby suggesting a less important role for the intrinsic chaperone ability of the Hsp40s.[48,49] However, future studies were able to show that the chaperone activity is indeed important for in vivo function by looking in yeast strains mutant for both Ydj1 and Sis1. In this genetic background, the substrate binding ability of either Ydj1 or Sis1 was necessary to maintain growth,[50] thereby suggesting an overlapping role of the Ydj1 and Sis1 chaperone activity. Therefore there must be unique roles besides the chaperone ability of the Sis1 protein, which are essential for viability. It is likely that there are unique roles of Sis1 that the Ydj1 protein cannot compensate for.

Determination of Specificity

Not only do Type I and Type II Hsp40s bind different substrates and have different chaperone activities in vitro, but they also have distinct functions in vivo. Sis1 is required to maintain cell growth and to maintain the prion state of the RNQ1 protein, while Ydj1 is not.[46,51] The unique function of Sis1 is conserved by the Type II Hsp40s since both human and Drosophila Type II Hsp40s can substitute for Sis1, while neither yeast nor human Type I Hsp40s can.[52] Sequence analysis reveals two possible regions that may be responsible for specifying this difference in function between the Type I and Type II Hsp40s. First the G/F rich region of Ydj1 and Sis1 are different, with that of Sis1 containing the 10 residue insert, GHAFSNEDAF.[52] Second, as mentioned previously, the protein modules located in the middle of Ydj1 and Sis1 are different such that Ydj1 contains the ZFLR and Sis1 contains the G/M region as well as a CTD1 domain (Fig. 2). Thus, it is plausible that either the G/F domain or the central domain (ZFLR vs. G/M CTD1) of Ydj1 and Sis1 serve to specify their in vivo functions. Below we will discuss the studies that were carried out in order to determine whether either of these differences has a role in specifying the functions of the Type I proteins versus the Type II proteins.

The G/F Region

To determine whether the G/F regions of Type I and Type II Hsp40s help specify Hsp70 functions the Craig group has carried out a number of complementation studies with Hsp40 fragments.[48] In these studies, which were conducted with a sis1Δ strain, the G/F region of Sis1, but not that of Ydj1, was shown to be important for suppression of lethality caused by the loss of Sis1 function.[48] Deletion of the G/F region also prevents Sis1 from maintaining the prion state of RNQ1, while truncated versions of Sis1 containing just the J domain and G/F region (Sis1 1-121) can functionally substitute for wild type Sis1.[51,52] In order to determine if the unique insertions of the G/F region allow the Sis1 G/F region to be functionally distinct from the Ydj1 G/F region, deletion analyses were carried out. Normally, truncated versions of Sis1 containing the J domain and G/F region (Sis1 1-121) are able to maintain cell growth in the absence of wild type protein. Deletion of one of the unique insertions of the Sis1 G/F region (Sis1 1-121 Δ101-113) causes a defect in cell growth in the absence of wild type Sis1, thereby suggesting that the unique insertion of the G/F region is at least partially responsible for specifying the in vivo functions of the Sis1 protein.[52]

Clues as to why the unique G/F region of Sis1 is important came from studies in which Sis1 ΔG/F and Ydj1 ΔG/F proteins were overexpressed in yeast.[53] The overexpression of Sis1 ΔG/F had deleterious effects on cell growth while the overexpression of Ydj1 ΔG/F did not. In order to determine why the G/F deletion had these deleterious effects, the Sis1 ΔG/F protein was purified and tested for its ability to function in different ways.[53] Sis1 ΔG/F could still bind denatured luciferase and the RNQ1 protein, and Sis1 ΔG/F could still stimulate the Hsp70 ATPase activity. The function that was lacking in the Sis 1 ΔG/F protein was the ability to cooperate with Hsp70 to refold denatured substrates. Since Sis 1 ΔG/F can still bind substrates and stimulate ATPase activity, the defect likely comes from an inability to efficiently transfer substrates from Sis1 to Hsp70. The combination of the genetic and biochemical studies discussed above suggests that the G/F region of the Type II Hsp40s may play a role in substrate transfer and thereby helps specify the unique functions of the Type II Hsp40s.

Central Domains

In addition to the differences found in the G/F regions, the central domains of the Type I and Type II Hsp40s also have dramatic structural differences. The central domain of the Type II Hsp40s contains the G/M region and a polypeptide binding site found in the CTD1, while the Type I Hsp40s contain a ZFLR. The differences in the substrate binding domains will be discussed in the next section so for now we will concentrate on how the G/M region versus the ZFLR may help specify function. Studies with the full length Sis1 protein indicate that the G/M region has some overlapping function with the G/F region.[52] As discussed above, deletion of unique residues within the G/F region has deleterious effects on cell growth in cells that only have a truncated version of Sis1 containing the J domain and G/F region. However, in cells expressing the full length Sis1, deletion of the same unique residues, Sis1 Δ101-113, no longer effects cell growth at normal temperatures. These cells also maintain the prion state of RNQ1. Likewise, deletion of the G/M region from the full length protein (Sis 1 ΔG/M) has no effect on cell growth at normal temperatures and has a very mild effect on the maintenance of the RNQ1 prion. However, deletion of both the G/M and the unique residues within the G/F region from the full length protein (Sis 1 ΔG/M Δ101-113) prevents the maintenance of the RNQ1 prion. These studies indicate that the essential function of Sis1 is actually specified by both the G/M region and the unique residues within the G/F region.[52]

Studies of the ZFLR of Type I Hsp40s have also provided clues as to why the function of the Type I proteins is unique from the Type II proteins. While the central domain of the Type I Hsp40s, the ZFLR, has been implicated as a component of the polypeptide binding site in combination with the adjacent C-terminal region for Type I Hsp40s,[32,54,55] the exact role of the ZFLR is not completely clear. A NMR structure of the ZFLR reveals a V-shaped groove with an extended β-hairpin topology, which could potentially be involved in protein:protein interactions.[56] However, Hsp40 ZFLR mutants which do exhibit defects in protein folding activity, do not exhibit defects in polypeptide binding.[32] In addition, deletion of the ZFLR from DnaJ does not abolish substrate binding.[55] A proteolytic fragment of Ydj1, Ydj1 (179-384), which is missing the J-domain and the first zinc binding module of the ZFLR is also capable of suppressing protein aggregation and therefore must retain the ability to bind substrates.[32] Therefore, while these studies do not rule out the possibility that the ZFLR is involved with polypeptide binding, it is definitely not required for polypeptide binding.

Mutation of the ZFLR does reveal that this domain is necessary to cooperate with Hsp70 in folding reactions.[32,57,58] In order to determine why the ZFLR is necessary to cooperate with Hsp70, yeast cells expressing a zinc binding domain 2 (ZBD2) mutant of Ydj1 were examined.[57] These cells show a decrease in the activity of the androgen receptor (AR), which is a known Hsp70 substrate. Isolation of androgen receptor complexes revealed that mutation of the ZFLR of Ydj1 leads to the accumulation of Hsp40-AR complexes with the concomitant decrease in Hsp70-AR complexes.[57] Therefore, it seems that one important role of the ZFLR is to stimulate the transfer of substrates from Hsp40 to Hsp70.

In order to directly decipher the involvement of the ZFLR versus the G/M CTD1 central domains in specifying Hsp40 function, chimeric forms of Ydj1 and Sis1 were constructed in which the central domains were swapped to form YSY and SYS.[44] Purified SYS and YSY were found to exhibit protein-folding activity and substrate specificity that mimicked that of Ydj1 and Sis1, respectively.[44] In vivo studies also showed that YSY exhibited a gain of function, and unlike Ydj1, could complement the lethal phenotype of sis1Δ and promote the propagation of the yeast prion [RNQ$^+$]. SYS exhibited a loss of function and was unable to maintain [RNQ$^+$]. These in vitro and in vivo data suggest that the central domain of Ydj1 and Sis1 are exchangeable and that they help specify Hsp40's cellular functions.[44]

Substrate Binding Domains

The studies discussed above suggest that the unique residues in the G/F region and the different central domains may help specify the function of the Type I vs. the Type II proteins by affecting the manner in which the individual chaperones interact with or transfer substrates to Hsp70. Another important determinant of specificity could obviously come from the substrate binding domains themselves. Since the Type I proteins do prefer to bind peptides that are distinct from those that the Type II proteins bind,[44] one would hypothesize that there are differences in the substrate binding domains of these two types of proteins. Studies have shown that the substrate binding domains of both Type I and Type II Hsp40s are found in the C-terminal domain.[32,33,36,54,55] For example, the carboxyl terminus of the Ydj1 protein (residues 206-380) was shown to be at least partially responsible for polypeptide binding,[32] and a single point mutation in this C-terminal domain (Ydj1 G315D) exhibits severe defects in polypeptide binding.[32] A fragment of Ydj1 consisting of residues 179-384 was also shown to be able to suppress rhodanese aggregation to the same level as the full length protein.[32] This fragment lacks the J domain, the G/F region and the first zinc binding domain, but contains the C-terminal domain. Studies of the yeast Type II Hsp40, Sis1, have also localized the polypeptide binding site to the C-terminal domain,[36] and a C-terminal fragment that is truncated at the end of the G/M region and contains the unique CTD1 domain as well as the CTD2 domain (Sis1 171-352) was shown to bind denatured luciferase as efficiently as the full length protein.[33] Yeast two hybrid analyses showed that the CTD1 or CTD2 domains of Hdj1 cannot function alone but instead must function together to bind the substrate, ankyrin-B.[40] Therefore, similar regions within Ydj1 and Sis1 are implicated in polypeptide binding.

Crystal structures of the C-terminal domains of both Ydj1 and Sis1 have been solved (Fig. 3).[36,59,60] These structures confirm that the C-terminal domain is a site for peptide binding for both types of Hsp40s and they suggest similar yet unique mechanisms for substrate binding. The Ydj1 crystal structure is of the monomer form of a truncated C-terminal domain (Ydj1 F335D 102-350) in complex with a short peptide substrate, GWLYEIS.[60] The crystal structure showed that the truncated C-terminal domain forms an L-shaped structure, which can be broken up into 3 domains. Domain two contains the ZFLR, while domains 1 and 3 each contain a hydrophobic depression (Fig. 3A). The crystal structure shows that the peptide substrate binds to Ydj1 by forming an extra β-strand in the domain 1 depression.[60] There is also an interaction in which the leucine from the peptide is buried in a small hydrophobic pocket found in this surface depression (Fig. 3A). The pocket that the leucine is buried in is formed by a variety of highly conserved hydrophobic residues (I116, Leu135, Leu 137, Leu 216, and Phe 249), thereby suggesting that the pocket may be a common feature found in Type I Hsp40s, and may play a role in determining the substrate specificity. Mutational studies were later carried out which verified that this hydrophobic pocket on the surface of the Ydj1 domain 1 is indeed required for peptide binding and for both Hsp70 dependent and independent chaperone activity.[61]

The second hydrophobic depression in domain 3 could potentially be a part of the substrate binding site, but the short length of the peptide in this study prevented the identification of any such interactions. The depression in domain 3 also had a small hydrophobic pocket, but in this case it was occupied by a unique phenylalanine found only in domain 3, but not domain 1. So, whether or not an extended polypeptide substrate would interact with both depressions on the Ydj1 monomer has yet to be determined. In order to determine how the individual Ydj1 monomers would line up with each other, the structure of the Ydj1 F335D 102-350 monomer was later combined with structural data of the Ydj1 dimerization domain (Ydj1 253-381) to form a more complete model of the Ydj1 molecule.[62] This model proposes that a large cleft is formed between two monomers and that the zinc finger like regions point directly towards one another, but are not in close proximity to the bound peptide substrate. These data reinforce the idea discussed above that the zinc finger like domains most likely play a role in specifying Type I function not by being directly involved in substrate binding, but instead by mediating productive interactions between Hsp40 and Hsp70 that allow for the transfer of substrate.

The X-ray crystal structure of Sis1 171-352 was also solved[36,63] and it depicts a homodimer that has a crystallographic two-fold axis (Fig. 3B). Sis1 171-352 monomers are elongated and constructed from two barrel-like domains that have similar folds and mostly β-structure. Sis1 dimerizes through a short C-terminal α-helical domain, and the dimer has a wishbone shape with a cleft that separates the arms of the two elongated monomers. The CTD1 on each monomer also contains two shallow depressions that are lined by highly conserved solvent exposed hydrophobic residues (Fig. 3B). Mutational analysis of the residues that line the hydrophobic depression in Sis1 has identified K199, F201 and F251 as amino acids that are essential for cell

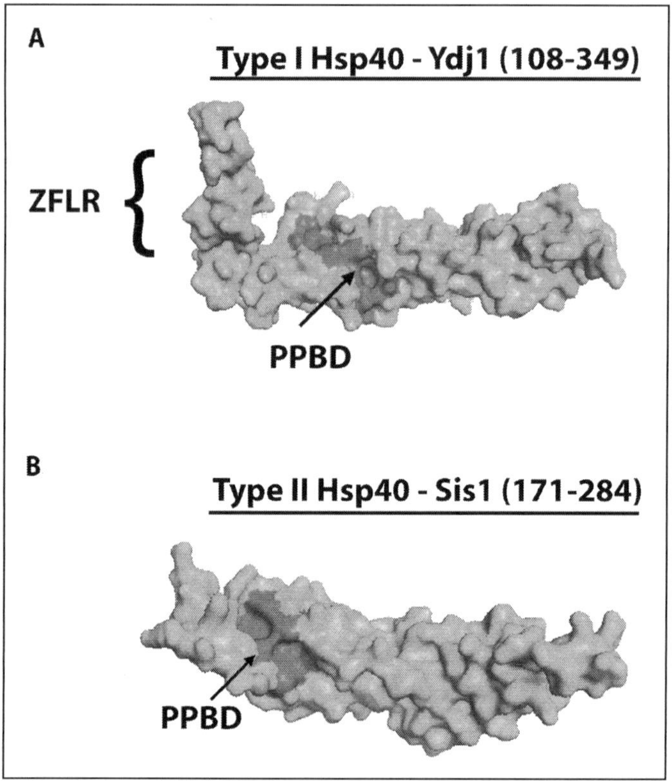

Figure 3. Structures of the Hsp40 peptide binding sites. Models of the Ydj1 and Sis1 C-terminal peptide binding sites were rendered from PDB files INLT and 1C3G respectively with the MacPyMol software package, and then labeled using Adobe Photoshop. A) The Ydj1 structure represents amino acids 108-349 from the structure that was solved of Ydj1 F335D (102-350) in complex with a peptide substrate.[60] In this figure, the peptide has been subtracted out to show the surface of the Ydj1 molecule that corresponds to the peptide binding domain. The red highlighted areas indicate two hydrophobic pockets. The lower pocket is where the leucine of the peptide substrate makes an interaction with Ydj1. The upper hydrophobic depression did not show any contacts with substrate perhaps due to the short length of the peptide substrate. B) A model of the monomer of Sis1 (171-284) that came from the crystal structure of Sis1 (171-352).[36] The surface shown in this figure depicts the contours on the CTD1 where a hydrophobic groove with two shallow depressions is visible. Hydrophobic residues are labeled in red. PPBD: polypeptide binding domain; ZFLR: zinc finger like region. A color version of this figure is available online at www.eurekah.com.

viability and required for Sis1 to both bind denatured substrates and cooperate with Hsp70 to refold those substrates.[35]

Hsp40 Quaternary Structure

A common feature of the Type I and Type II Hsp40s is that dimerization was shown to be necessary for the chaperone activity of both.[37,64] The residues responsible for dimerization are found in the C-termini of both types of proteins.[36,62,64] Deletion of the C-terminal (331-376) residues of the bacterial Type I Hsp40, DnaJ, not only prevented dimerization, but it also decreased the affinity for peptide substrate and destroyed the ability of the DnaJ to prevent aggregation of GAPDH.[64] Interestingly, monomerization of the yeast Type I protein, Ydj1, does not prevent binding of short peptide substrates,[60] but how the monomerization affects the affinity for a polypeptide substrate or its chaperone activity is unknown. Monomerization of the human Type I Hsp40, DjA1 (Hdj-2), also had a deleterious affect on binding of denatured luciferase.[37] Dimerization is also a requirement for the Type II Hsp40, Sis1, to maintain denatured luciferase in a folding competent state and cooperate with Hsp70 to refold denatured luciferase.[36] Whether or not dimerization is required for simple polypeptide substrate binding by Sis1, however, is not known. Since disruption of dimerization does affect the ability of the Type I Hsp40, Hdj-2, to bind substrate,[37] it would be interesting to determine if the two binding sites of Sis1 (one on each monomer) are also necessary for the Type II proteins to bind substrate. While the Type I and Type II proteins seem to behave similarly when monomerization is induced, crystal structures show that the mechanism by which the dimerization occurs in each type is actually very different.[62] The Type I protein, Ydj1, utilizes a unique mechanism whereby a β-sheet is formed between the dimerization motif and domain III of the protein. Whether or not this unique dimerization motif plays a role in specifying the function of the Type I proteins has yet to be determined.

The unique domain structure of the different types of Hsp40 proteins could also contribute to unique quaternary structures, and any differences in overall structure of the proteins could easily specify different functions for the proteins. Recent studies have used small angle X-ray scattering (SAXS) to determine the quaternary structure of the human Type I Hsp40, DjA1, and the human type II Hsp40, DjB4.[37] The SAXS data does not provide as high resolution data as the X-ray crystallography or NMR structures discussed previously, but this technique does allow for the determination of density and overall protein shape in solution. Therefore, the combination of the NMR structural data and the X-ray crystallography data with the SAXS data provides a very powerful tool to help determine exactly how the protein domains are oriented in solution. The SAXS data collected from the human Type I and Type II proteins[37] were modeled together with the crystal structure data from the yeast Ydj1 and Sis1 proteins[36,60] as well as the NMR data of the J domain.[65] These studies show that there are indeed substantial differences in the quaternary structure of the Type I and Type II Hsp40 that may help account for the different substrate specificity and chaperone activity (Fig. 4).[37]

The models (Fig. 4),[37] show that the Type I Hsp40 forms a bullet shaped compact dimer with both the N- and C-termini of each monomer facing the opposing N- or C-termini of the other monomer respectively. The ZFLR of each monomer directly face each other, and may act to ensure proper spacing between the two monomers. The Type II Hsp40, DjB4, forms a more extended dimer where only the C-termini of the opposing monomers interact. This model helps us understand why the central domains of the Type I and Type II regions may help specify function. From the model it looks as if the ZFLR may be responsible for holding the two individual monomers at an appropriate distance from each other. This may allow for productive and unique interactions with both substrates and/or Hsp70. The extension of the G/F region and the addition of the G/M region in the Type II Hsp40 allows for a different structure in which the J domains are extended out away from each other. Based on the structural data in this study, along with previous data on DnaJ-DnaK and Sis1-Ssa1 interactions,[22,66-68] the authors have also proposed a model in which Type I Hsp40s and Type II Hsp40s use different mechanisms to interact with Hsp70s.[37]

While it appears that the unique structures almost certainly do play a role in specifying function, the exact mechanism by which these structures affect both the chaperone and co-chaperone function of the Hsp40s is not yet known. A combination of all the unique characteristics of the Type I and Type II chaperones discussed above likely explain the different levels of chaperone and co-chaperone activity that we see from the different types of Hsp40s. The unique domains and overall structures of the Hsp40s may also allow for certain Hsp40s to be localized to specific multi-chaperone complexes where they can play individualized roles.

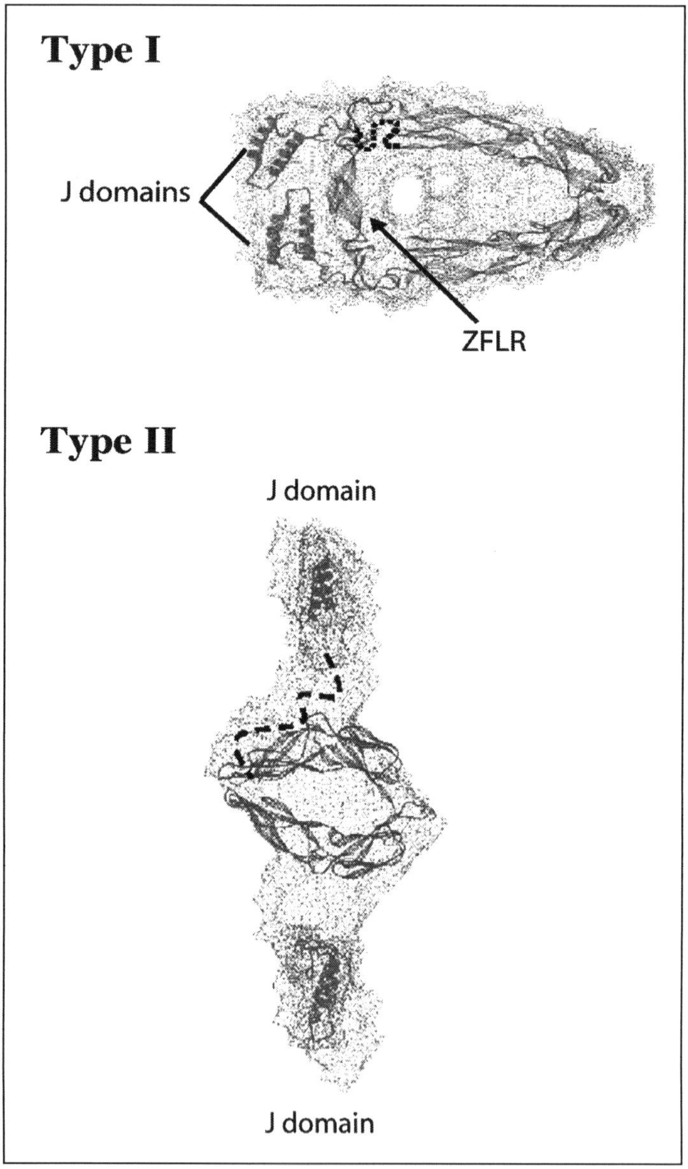

Figure 4. Models of Type I and Type II quaternary structures. Ab initio models of Type I and Type II Hsp40s were generated by SAXS data and molecular modeling of the human DjA1 and DjB4 proteins. The dotted lines represent the G/F rich regions, and the J domains are labeled. The ZFLR corresponds to the zinc finger like region. These models show the major differences in the quaternary structure of the Type I and Type II Hsp40s. The Adobe Photoshop software was used to render this figure. Reprinted from Borges JC et al, J Biol Chem 2005;280(14):13671-13681;[37] ©2005, with permission from American Society for Biochemistry and Molecular Biology in the format Other Book via Copyright Clearance Center. A color version of this figure is available online at www.eurekah.com.

Acknowledgements

Work in the laboratory of DMC is supported by the National Institutes of Health. MFNR is supported by a SPIRE postdoctoral fellowship grant through the National Institute of General Medical Sciences (NIGMS). Figure 4 was reprinted from the Journal of Biological Chemistry by Ramos, Carlos; ©2005 with permission from American Society for Biochemistry and Molecular Biology in the format Other Book via Copyright Clearance Center.

References

1. Hartl FU, Hayer-Hartl M. Molecular chaperones in the cytosol: From nascent chain to folded protein. Science 2002; 295(5561):1852-1858.
2. Terada K, Kanazawa M, Bukau B et al. The human DnaJ homologue dj2 facilitates mitochondrial protein import and luciferase refolding. J Cell Biol 1997; 139(5):1089-1095.
3. Ngosuwan J, Wang NM, Fung KL et al. Roles of cytosolic Hsp70 and Hsp40 molecular chaperones in post-translational translocation of presecretory proteins into the endoplasmic reticulum. J Biol Chem 2003; 278(9):7034-7042.
4. Abdul KM, Terada K, Gotoh T et al. Characterization and functional analysis of a heart-enriched DnaJ/ Hsp40 homolog dj4/DjA4. Cell Stress Chaperones 2002; 7(2):156-166.
5. Prip-Buus C, Westerman B, Schmitt M et al. Role of the mitochondrial DnaJ homologue, Mdj1p, in the prevention of heat-induced protein aggregation. FEBS Lett 1996; 380(1-2):142-146.
6. Schroder H, Langer T, Hartl FU et al. DnaK, DnaJ and GrpE form a cellular chaperone machinery capable of repairing heat-induced protein damage. EMBO J 1993; 12(11):4137-4144.
7. Wacker JL, Zareie MH, Fong H et al. Hsp70 and Hsp40 attenuate formation of spherical and annular polyglutamine oligomers by partitioning monomer. Nat Struct Mol Biol 2004; 11(12):1215-1222.
8. Fink AL. Chaperone-mediated protein folding. Physiol Rev 1999; 79(2):425-449.
9. Walsh P, Bursac D, Law YC et al. The J-protein family: Modulating protein assembly, disassembly and translocation. EMBO Rep 2004; 5(6):567-571.
10. Venter JC, Adams MD, Myers EW et al. The sequence of the human genome. Science 2001; 291(5507):1304-1351.
11. Cheetham ME, Caplan AJ. Structure, function and evolution of DnaJ: Conservation and adaptation of chaperone function. Cell Stress Chaperones 1998; 3(1):28-36.
12. Kelley WL. The J-domain family and the recruitment of chaperone power. Trends Biochem Sci 1998; 23(6):222-227.
13. Lemmon SK. Clathrin uncoating: Auxilin comes to life. Curr Biol 2001; 11(2):R49-52.
14. Fan CY, Lee S, Cyr DM. Mechanisms for regulation of Hsp70 function by Hsp40. Cell Stress Chaperones 2003; 8(4):309-316.
15. Mayer MP, Bukau B. Hsp70 chaperones: Cellular functions and molecular mechanism. Cell Mol Life Sci 2005; 62(6):670-684.
16. Mayer MP, Bukau B. Hsp70 chaperone systems: Diversity of cellular functions and mechanism of action. Biol Chem 1998; 379(3):261-268.
17. Esser C, Alberti S, Hohfeld J. Cooperation of molecular chaperones with the ubiquitin/proteasome system. Biochim Biophys Acta 2004; 1695(1-3):171-188.
18. Hohfeld J, Minami Y, Hartl FU. Hip, a novel co-chaperone involved in the eukaryotic Hsc70/ Hsp40 reaction cycle. Cell 1995; 83(4):589-598.
19. Frydman J, Hohfeld J. Chaperones get in touch: The Hip-Hop connection. Trends Biochem Sci 1997; 22(3):87-92.
20. Lee DH, Sherman MY, Goldberg AL. Involvement of the molecular chaperone Ydj1 in the ubiquitin-dependent degradation of short-lived and abnormal proteins in Saccharomyces cerevisiae. Mol Cell Biol 1996; 16(9):4773-4781.
21. Youker RT, Walsh P, Beilharz T et al. Distinct roles for the Hsp40 and Hsp90 molecular chaperones during cystic fibrosis transmembrane conductance regulator degradation in yeast. Mol Biol Cell 2004; 15(11):4787-4797.
22. Laufen T, Mayer MP, Beisel C et al. Mechanism of regulation of hsp70 chaperones by DnaJ co-chaperones. Proc Natl Acad Sci USA 1999; 96(10):5452-5457.
23. Hennessy F, Nicoll WS, Zimmermann R et al. Not all J domains are created equal: Implications for the specificity of Hsp40-Hsp70 interactions. Protein Sci 2005; 14(7):1697-1709.
24. Kelley WL. Molecular chaperones: How J domains turn on Hsp70s. Curr Biol 1999; 9(8):R305-308.
25. Wickner S, Hoskins J, McKenney K. Function of DnaJ and DnaK as chaperones in origin-specific DNA binding by RepA. Nature 1991; 350(6314):165-167.

26. Langer T, Lu C, Echols H et al. Successive action of DnaK, DnaJ and GroEL along the pathway of chaperone-mediated protein folding. Nature 1992; 356(6371):683-689.
27. Hendrick JP, Langer T, Davis TA et al. Control of folding and membrane translocation by binding of the chaperone DnaJ to nascent polypeptides. Proc Natl Acad Sci USA 1993; 90(21):10216-10220.
28. Szabo A, Langer T, Schroder H et al. The ATP hydrolysis-dependent reaction cycle of the Escherichia coli Hsp70 system DnaK, DnaJ, and GrpE. Proc Natl Acad Sci USA 1994; 91(22):10345-10349.
29. Meacham GC, Lu Z, King S et al. The Hdj-2/Hsc70 chaperone pair facilitates early steps in CFTR biogenesis. EMBO J 1999; 18(6):1492-1505.
30. Shen Y, Hendershot LM. ERdj3, a stress-inducible endoplasmic reticulum DnaJ homologue, serves as a cofactor for BiP's interactions with unfolded substrates. Mol Biol Cell 2005; 16(1):40-50.
31. Cyr DM. Cooperation of the molecular chaperone Ydj1 with specific Hsp70 homologs to suppress protein aggregation. FEBS Lett 1995; 359(2-3):129-132.
32. Lu Z, Cyr DM. The conserved carboxyl terminus and zinc finger-like domain of the co-chaperone Ydj1 assist Hsp70 in protein folding. J Biol Chem 1998; 273(10):5970-5978.
33. Lu Z, Cyr DM. Protein folding activity of Hsp70 is modified differentially by the Hsp40 co-chaperones Sis1 and Ydj1. J Biol Chem 1998; 273(43):27824-27830.
34. Han W, Christen P. Mechanism of the targeting action of DnaJ in the DnaK molecular chaperone system. J Biol Chem 2003; 278(21):19038-19043.
35. Lee S, Fan CY, Younger JM et al. Identification of essential residues in the type II Hsp40 Sis1 that function in polypeptide binding. J Biol Chem 2002; 277(24):21675-21682.
36. Sha B, Lee S, Cyr DM. The crystal structure of the peptide-binding fragment from the yeast Hsp40 protein Sis1. Structure Fold Des 2000; 8(8):799-807.
37. Borges JC, Fischer H, Craievich AF et al. Low resolution structural study of two human HSP40 chaperones in solution. DJA1 from subfamily A and DJB4 from subfamily B have different quaternary structures. J Biol Chem 2005; 280(14):13671-13681.
38. Minami Y, Hohfeld J, Ohtsuka K et al. Regulation of the heat-shock protein 70 reaction cycle by the mammalian DnaJ homolog, Hsp40. J Biol Chem 1996; 271(32):19617-19624.
39. Freeman BC, Morimoto RI. The human cytosolic molecular chaperones hsp90, hsp70 (hsc70) and hdj-1 have distinct roles in recognition of a nonnative protein and protein refolding. EMBO J 1996; 15(12):2969-2979.
40. Mohler PJ, Hoffman JA, Davis JQ et al. Isoform specificity among ankyrins. An amphipathic alpha-helix in the divergent regulatory domain of ankyrin-b interacts with the molecular co-chaperone Hdj1/Hsp40. J Biol Chem 2004; 279(24):25798-25804.
41. Bischofberger P, Han W, Feifel B et al. D-Peptides as inhibitors of the DnaK/DnaJ/GrpE chaperone system. J Biol Chem 2003; 278(21):19044-19047.
42. Rudiger S, Schneider-Mergener J, Bukau B. Its substrate specificity characterizes the DnaJ co-chaperone as a scanning factor for the DnaK chaperone. EMBO J 2001; 20(5):1042-1050.
43. Zhu X, Zhao X, Burkholder WF et al. Structural analysis of substrate binding by the molecular chaperone DnaK. Science 1996; 272(5268):1606-1614.
44. Fan CY, Lee S, Ren HY et al. Exchangeable chaperone modules contribute to specification of type I and type II Hsp40 cellular function. Mol Biol Cell 2004; 15(2):761-773.
45. Freeman BC, Myers MP, Schumacher R et al. Identification of a regulatory motif in Hsp70 that affects ATPase activity, substrate binding and interaction with HDJ-1. EMBO J 1995; 14(10):2281-2292.
46. Luke MM, Sutton A, Arndt KT. Characterization of SIS1, a Saccharomyces cerevisiae homologue of bacterial dnaJ proteins. J Cell Biol 1991; 114(4):623-638.
47. Caplan AJ, Douglas MG. Characterization of YDJ1: A yeast homologue of the bacterial dnaJ protein. J Cell Biol 1991; 114(4):609-621.
48. Yan W, Craig EA. The glycine-phenylalanine-rich region determines the specificity of the yeast Hsp40 Sis1. Mol Cell Biol 1999; 19(11):7751-7758.
49. Johnson JL, Craig EA. A role for the Hsp40 Ydj1 in repression of basal steroid receptor activity in yeast. Mol Cell Biol 2000; 20(9):3027-3036.
50. Johnson JL, Craig EA. An essential role for the substrate-binding region of Hsp40s in Saccharomyces cerevisiae. J Cell Biol 2001; 152(4):851-856.
51. Sondheimer N, Lopez N, Craig EA et al. The role of Sis1 in the maintenance of the [RNQ+] prion. EMBO J 2001; 20(10):2435-2442.
52. Lopez N, Aron R, Craig EA. Specificity of class II Hsp40 Sis1 in maintenance of yeast prion [RNQ+]. Mol Biol Cell 2003; 14(3):1172-1181.
53. Aron R, Lopez N, Walter W et al. In vivo bipartite interaction between the Hsp40 Sis1 and Hsp70 in Saccharomyces cerevisiae. Genetics 2005; 169(4):1873-1882.

54. Szabo A, Korszun R, Hartl FU et al. A zinc finger-like domain of the molecular chaperone DnaJ is involved in binding to denatured protein substrates. EMBO J 1996; 15(2):408-417.
55. Banecki B, Liberek K, Wall D et al. Structurefunction analysis of the zinc finger region of the DnaJ molecular chaperone. J Biol Chem 1996; 271(25):14840-14848.
56. Martinez-Yamout M, Legge GB, Zhang O et al. Solution structure of the cysteine-rich domain of the Escherichia coli chaperone protein DnaJ. J Mol Biol 2000; 300(4):805-818.
57. Fan CY, Ren HY, Lee P et al. The type I Hsp40 zinc finger-like region is required for Hsp70 to capture nonnative polypeptides from Ydj1. J Biol Chem 2005; 280(1):695-702.
58. Linke K, Wolfram T, Bussemer J et al. The roles of the two zinc binding sites in DnaJ. J Biol Chem 2003; 278(45):44457-44466.
59. Li J, Sha B. Preliminary X-ray crystallographic studies of yeast Hsp40 Ydj1 complexed with its peptide substrate. Acta Crystallogr D Biol Crystallogr 2003; 59(7):1317-1319.
60. Li J, Qian X, Sha B. The crystal structure of the yeast Hsp40 Ydj1 complexed with its peptide substrate. Structure (Camb) 2003; 11(12):1475-1483.
61. Li J, Sha B. Structurebased mutagenesis studies of the peptide substrate binding fragment of type I heat-shock protein 40. Biochem J 2005; 386(3):453-460.
62. Wu Y, Li J, Jin Z et al. The crystal structure of the C-terminal fragment of yeast Hsp40 Ydj1 reveals novel dimerization motif for Hsp40. J Mol Biol 2005; 346(4):1005-1011.
63. Sha B, Cyr D. Purification, crystallization and preliminary X-ray crystallographic studies of S. cerevisiae Hsp40 Sis1. Acta Crystallogr D Biol Crystallogr 1999; 55(6):1234-1236.
64. Shi YY, Hong XG, Wang CC. The C-terminal (331-376) sequence of Escherichia coli DnaJ is essential for dimerization and chaperone activity: A small angle X-ray scattering study in solution. J Biol Chem 2005; 280(24):22761-22768.
65. Qian YQ, Patel D, Hartl FU et al. Nuclear magnetic resonance solution structure of the human Hsp40 (HDJ-1) J-domain. J Mol Biol 1996; 260(2):224-235.
66. Suh WC, Burkholder WF, Lu CZ et al. Interaction of the Hsp70 molecular chaperone, DnaK, with its co-chaperone DnaJ. Proc Natl Acad Sci USA 1998; 95(26):15223-15228.
67. Gassler CS, Buchberger A, Laufen T et al. Mutations in the DnaK chaperone affecting interaction with the DnaJ co-chaperone. Proc Natl Acad Sci USA 1998; 95(26):15229-15234.
68. Qian X, Hou W, Zhengang L et al. Direct interactions between molecular chaperones heat-shock protein (Hsp) 70 and Hsp40: Yeast Hsp70 Ssa1 binds the extreme C-terminal region of yeast Hsp40 Sis1. Biochem J 2002; 361(1):27-34.

Chapter 5

The Chaperone and Co-Chaperone Activities of Cdc37 during Protein Kinase Maturation

Avrom J. Caplan*

Abstract

Eukaryotic protein kinases fold in the cytosol in association with the Hsp90 molecular chaperone machine. This machine comprises a large number of chaperones and co-chaperones, among them Cdc37, which is essential for protein kinase maturation. Cdc37 interacts with protein kinases via their catalytic domains and with Hsp90. Cdc37 inhibits Hsp90's ATPase activity which is thought to promote assembly of the misfolded kinase into a multi-chaperone complex. This function shows that Cdc37 is a co-chaperone of Hsp90. However, Cdc37 also has chaperone activity of its own and can protect protein kinases against unfolding and degradation. Recent studies suggest a model in which the chaperone activities of Hsp90 and Cdc37 cooperate to promote efficient protein kinase folding.

Introduction

Cdc37 is a molecular chaperone that is required for folding of protein kinases. This chapter will focus on the issue of how Cdc37 functions as a molecular chaperone and as a co-chaperone of the larger Hsp90 machinery. Its ability to function in both capacities derives from Cdc37's ability to interact with misfolded protein kinases as well as with Hsp90 itself. While these binding activities are distinct, it is thought that they allow for coordinated action of both Cdc37 and Hsp90 in the folding process. Several previous reviews have focused on Cdc37 involvement in protein kinase folding, and the reader is referred to these for background information.[1-3] In this chapter, I will focus on what is known about the contribution of Cdc37 as a chaperone and as a co-chaperone to the folding process.

Cdc37 was discovered in yeast as a gene required for cell cycle progression (hence the *cdc* designation for cell division cycle).[4,5] Further studies noted genetic interactions between Cdc37 and the yeast cyclin-dependent kinase, Cdc28, and that Cdc37 was important for Cdc28 stability.[6,7] Meanwhile, biochemical analyses of Hsp90 tertiary organization showed that it could complex with a protein called p50, and that Hsp90/p50 could be found in complexes with protein kinases.[8,9] Subsequent cloning of the gene for p50 found it to be identical with Cdc37.[10,11] Continued interest in Cdc37 derives from findings that it is upregulated in tumor cells[12] and is required for cell proliferation.[13] As such, Cdc37 presents an excellent target for chemotherapy in a similar manner to the way in which Hsp90 has been targeted using benzoquinoid ansamycins.

Cdc37 interacts with many protein kinases and a few nonkinase clients, such as androgen receptor, a viral reverse transcriptase and MyoD.[14-16] Its function in biogenesis of these nonkinase

*Avrom J. Caplan—Department of Pharmacology and Biological Chemistry, Box 1603, Mount Sinai School of Medicine, One Gustave L. Levy Place, New York, New York 10029, U.S.A. Email: avrom.caplan@mssm.edu

Networking of Chaperones by Co-Chaperones, edited by Gregory L. Blatch.

clients is barely understood, nor is it known which domains of the chaperone interact with them. By contrast, an increasingly sophisticated picture of Cdc37's interaction with protein kinases is emerging. Cdc37 interacts with both serine/threonine and tyrosine specific protein kinases although one of the many questions that remain to be answered is how many kinases need Cdc37 for activity? Based on recent analyses of Cdc37:kinase interactions it seems likely that the chaperone will have a general role in kinome biogenesis since Cdc37 interacts with highly conserved sequence motifs found in almost all kinases (more details below). Details of the kinases currently known to interact with Cdc37 may be found at the following web site: www.picard.ch/downloads/Cdc37interactors.pdf.

The distinction between 'chaperone' and 'co-chaperone' actions is currently not very clear.[17] For the purpose of this review, the chaperone activity reflects the ability of Cdc37 to interact with misfolded client polypeptides, protecting them against aggregation and degradation. This binding action may also involve promoting the folding event itself but this is not necessary for a protein to be defined as a chaperone. The co-chaperone action reflects the ability of Cdc37 to interact with other chaperones and modulate their activity. In this case, Cdc37 is an integral part of the Hsp90 chaperone machine, and is a regulator of the ATPase activity of Hsp90. Indeed, there are two quite different views in the literature of how Cdc37 functions. One is that Cdc37 is a kinase-targeting subunit of the Hsp90 chaperone machine.[3,10] In this case, it is proposed that the main function of Cdc37 is to interact with protein kinases and then target them to Hsp90 for proper folding to the active state. In this hypothesis, Cdc37 may play a passive role. In other studies (see below), there is a suggestion that Cdc37 may play a more active role in the folding process rather than passively transferring misfolded kinases to Hsp90.

Cdc37 Structure

Mammalian Cdc37 is 44.5 kDa, exists in a dimeric form and comprises three distinct domains (Fig. 1). The gene is known to be essential in several organisms and Cdc37 protein has a cytosolic localization. The N-domain consists of the first 126 amino acids and binds to protein kinases (see below). The N-domain structure is unknown although it contains the most conserved sequences among Cdc37s from different phyla. This conservation is highest over the first 30 amino acids and contains residues that confer kinase binding ability. Cdc37 from yeast has a large insertion in the N-domain compared with Cdc37s from metazoan phyla, and is larger than its mammalian paralogs at 58.4 kDa. The middle domain, by contrast, is quite well characterized from a structural perspective. It comprises a protease stable α-helical structure of approximately 150 amino acids that binds to Hsp90.[18-20] The crystal structure of this domain reveals a compact globular 6-helix bundle of approximately 80 amino acids that interacts with the N-terminal ATP binding domain of Hsp90. C-terminal to this bundle is a long α-helix that appears to separate the middle domain from the rest of the C-terminal portion of Cdc37.[18] The proximal sequences of this long helix are part of the protease-stable domain of 150 amino acids and technically belongs to the middle domain. The dimerization interface is thought to reside in the middle domain of Cdc37 as judged from studies with Cdc37 itself and by comparison to a related protein called Harc. Harc has a conserved middle domain with Cdc37 but a quite different N-domain architecture and does not bind to protein kinases.[21] The function of the C-domain of Cdc37 is not clear and is dispensable for growth at normal and stressful growth temperatures in *S. cerevisiae* and *S. pombe*.[22,23]

Cdc37 from both yeast and mammalian sources is phosphorylated by casein kinase II. Yeast Cdc37 is phosphorylated on both Ser14 and Ser17, while mammalian Cdc37 is phosphorylated on Ser13 (equivalent of Ser14 of yeast Cdc37; there is no Ser17 equivalent in the mammalian protein). Phosphorylation at Ser13 or Ser14 is important for protein kinase binding by Cdc37 and in its absence protein kinase maturation is inhibited, often leading to rapid proteasome-dependent degradation of the misfolded client.[24-26] The phosphorylation of Cdc37 is an intriguing and important finding, since casein kinase II requires Cdc37 for activity. This implies the existence of a positive feedback loop, whereby casein kinase II controls Cdc37 activity and therefore kinase folding on a broad scale.[24] What is unknown is whether any

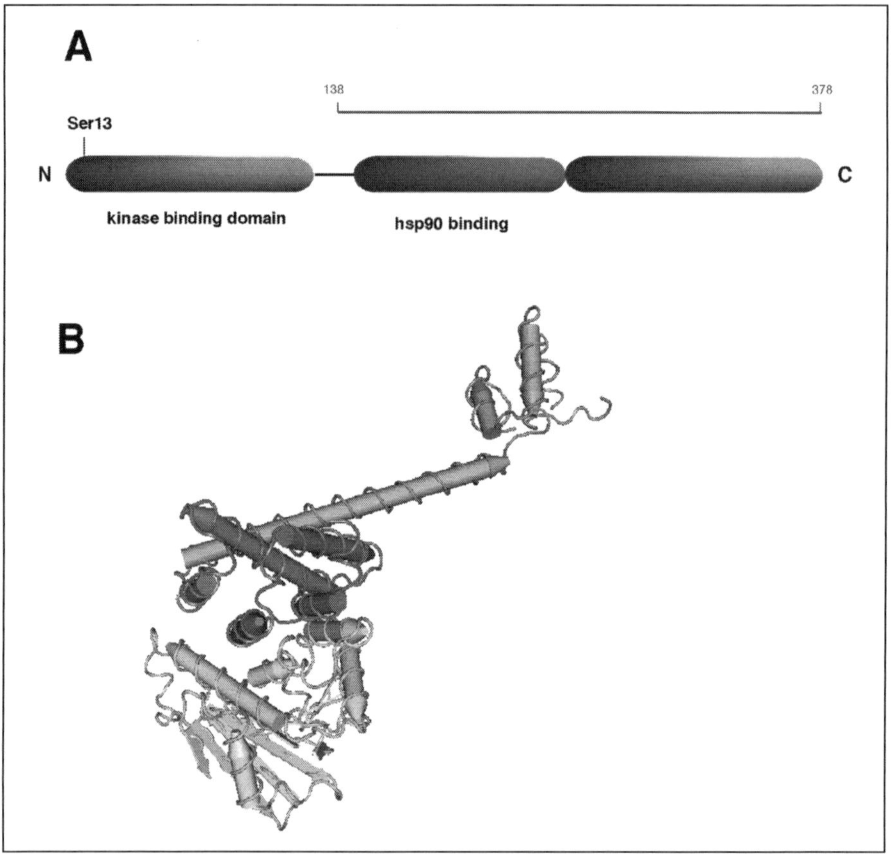

Figure 1. Domain organization and structure of Cdc37. A) Domain organization of Cdc37. Blue line denotes residues 138-378 which represent the limits of the structure described by Roe et al.[18] B) Structure of Cdc37 amino acids 138-378 with the isolated N-domain of Hsp90 (green). The compact 6-helix bundle that interacts with Hsp90 is shown in blue and the rest of the C-domain of Cdc37 shown in orange (adapted from ref. 18; PDB 1US7). Cn3D was used to render the structure (http://www.biosino.org/mirror/www.ncbi.nlm.nih.gov/Structure/cn3d/cn3d.html). A color version of this figure is available online at http://www.eurekah.com.

physiologically relevant dephosphorylation occurs, which might serve to inhibit folding at a kinomic level. Furthermore, the function of Cdc37 phoshorylation itself is unclear. One possibility is that it may serve to mimic one of the phosphates of ATP and provide a structural role during the chaperoning process. However, addition of ATP to immunoprecipitated Cdc37:Cdk4 complexes did not induce any disassembly.[27]

Cdc37 as a Co-Chaperone of Hsp90

Initial biochemical characterization of Cdc37, or p50 as its was first described, was as an Hsp90 binding protein in association with the viral oncogenic protein kinase v-Src.[28] Subsequent cloning of the gene for mammalian Cdc37 led to the first description of it as a kinase-targeting subunit of the larger Hsp90 chaperone machinery. This was based on the finding that Cdc37overexpression in insect cells led to a substantial increase in the amount of Hsp90 bound to Cdk4.[10] Subsequent studies have defined the interaction of Hsp90 with Cdc37 from both yeast and animal cells.

Mammalian Cdc37 interacts with Hsp90 with a dissociation constant of 1.5-2.5 μM[19,29] in the absence of a client kinase. By contrast yeast Cdc37 interaction with Hsp90 is much weaker having a dissociation constant of approximately 100 μM.[29] Both mammalian and yeast Cdc37 can inhibit the ATPase of Hsp90, although to quite different extents and in a manner that is consistent with the difference in binding affinities; i.e., mammalian Cdc37 is a much more potent inhibitor of Hsp90's ATPase than yeast Cdc37.[29] To a certain extent this may be due to the finding that Arg167 of mammalian Cdc37 points into the ATP binding pocket of Hsp90 and interacts with Glu33, which is thought to function in the ATPase reaction.[18] This interaction is therefore thought to play a role in inhibition of Hsp90's ATPase by Cdc37. By contrast, yeast Cdc37 has isoleucine at the equivalent position and is unlikely to regulate Hsp90's ATPase by a similar mechanism.

It is important to note that the crystal structure described by Roe et al[18] (see Fig. 1) provides only partial view of Hsp90:Cdc37 interaction. This is almost certainly because the kinase-binding N-domain of Cdc37 was not present in the complex containing the isolated N-domain of Hsp90. Indeed, full-length Cdc37 failed to interact with the isolated N-domain of Hsp90 suggesting that the kinase-binding domain of Cdc37 influences Hsp90:Cdc37 interactions.[19] This is supported by the finding that mutation of the conserved Trp7 near the N-terminus of Cdc37 led to decreased complex formation with Hsp90 present in cell lysates.[20] The N-domain and flexible linker regions of Hsp90 comprise the minimum amount needed for stable complex formation with full-length Cdc37. Furthermore, complexes were even more stable when the N-domain, flexible linker, and middle domains of Hsp90 were present (see ref. 30 for more information on Hsp90 structure). On the other hand, the same study found that Cdc37 truncations lacking the N-domain formed stable complexes with the isolated N-domain of Hsp90.[19] Whether the N-terminus of Cdc37 can interact with Hsp90, perhaps in the flexible linker is not clear, and no direct evidence for such an interaction has been presented. Alternatively, it is possible that the N-terminus of Cdc37 negatively influences the conformation of the middle domain that interacts with Hsp90.

Hsp90 interacts with a great many co-chaperones that form large complexes.[30,31] Cdc37's ability to interact with Hsp90 is coincident with the binding of some of the other co-chaperones but not all of them.[32] Aha1, for example, is a co-chaperone that stimulates Hsp90's ATPase, and is known to bind to the middle domain of Hsp90.[33,34] However, Cdc37 and Aha1 compete for binding to the same Hsp90 molecule, consistent with their roles at different stages of the Hsp90 cycle.[35] Interestingly, Cdc37 lacking the N-terminal kinase-binding domain does not compete with Aha1 for binding to Hsp90,[36] suggesting that N-terminal sequences in Cdc37 either bind to Hsp90, or otherwise sterically hinder Aha1 binding. In other studies, it was found that Cdc37 can coexist in complexes formed between Hsp90 and co-chaperone p23.[32,35] However, Cdc37 lacking its N-terminal kinase binding domain was unable to enter these complexes, once again suggesting a role for this domain in modulating the interaction between Cdc37 and Hsp90.[36]

Inhibition of Hsp90's ATPase by Cdc37 is thought to reflect an early stage in the cycle of Hsp90 action in client maturation. Another co-chaperone called Hop (Hsp70/Hsp90 organizing protein; Sti1 in yeast) also inhibits Hsp90's ATPase and is known to function in the loading of misfolded clients onto Hsp90 from complexes containing Hsp70.[37] Hop/Sti1 and yeast Cdc37 interact directly with each other. Moreover, Sti1 is required to stabilize Cdc37 in kinase client complexes with Hsp90 in yeast.[35,38,39] Deletion of *STI1* in yeast leads to defects in v-Src maturation and in signaling via a MAP kinase signaling pathway. In both cases, the defect resulting from *STI1* deletion was suppressed by *CDC37* overexpression. Furthermore, overexprexpressed *CDC37* helped recover stable binding of Hsp90 to a kinase in the *sti1Δ* mutant.[39] These combined data fully support the general model that Cdc37 functions in the loading of clients on Hsp90 and that it does so in association with Sti1. The ability of Cdc37 to function in place of Sti1 demonstrates that each protein has a similar role in the assembly process. However, it seems unlikely that they need to function together at all times for two reasons. The first is that *STI1* is not an essential gene whereas *CDC37* is essential; indeed, deletion of *STI1* does not result in a growth phenotype under normal growth conditions.[40] The second is that although we have found evidence for a role of Sti1 in maturation of some yeast kinases, we also found some kinases that need Cdc37 but not Sti1 for their maturation to the active state (Lee, P. and Caplan, A.J., unpublished observations).

The combined actions of Cdc37 and Sti1 of inhibiting Hsp90's ATPase appear to arrest the cycle for the purpose of client loading. Also, the binding of Hsp90 together with Cdc37 and the client kinase forms salt-stable complexes that can be distinguished from more labile complexes that dissociate at lower salt concentrations.[32] Once this is achieved, the cycle may resume and Hsp90's ATPase may be stimulated by the actions of Aha1, and this somehow promotes client folding and/or dissociation of the complex in a manner that is also facilitated by the co-chaperone p23.[41] This view points to a rather passive role for Cdc37 as a protein that simply loads clients onto Hsp90. Genetic studies, however, are beginning to paint a more complex portrait of Cdc37 as a chaperone that may function independently of Hsp90 in client kinase folding and maturation.

Chaperone Actions of Cdc37 in Protein Kinase Maturation

Initial characterization of Cdc37 as a molecular chaperone demonstrated that it could maintain β-galactosidase in a folding-competent conformation.[42] Cdc37 could not function to refold β-galactosidase itself—suggesting it had 'holdase' activity but not 'foldase' activity. Furthermore, purified Cdc37 prevented salt induced denaturation of a client kinase in vitro.[42]

Interaction of Cdc37 with client kinases has been investigated by mutagenesis and truncation studies in an attempt to identify the sites of interaction. Eukaryotic protein kinases have a highly conserved bi-lobal architecture. The N-domain is primarily β-sheet while the C-domain is almost entirely α-helical. ATP fits in a deep groove between the lobes and is stabilized by residues in the N-domain. Previous studies noted that Cdc37 interacted with the catalytic domain of different kinases.[43,44] Further analyses led to the conclusion that Cdc37 binding was restricted to sequences in the N-domain, which in unliganded protein kinase A is rather unstable even in the folded molecule.[45] Truncation analysis of Cdk4 revealed that two conserved glycines in the glycine-rich loop serve to stabilize kinase:Cdc37 interactions. However, sequences that extended from the glycine-rich loop to the loop between the α-C helix and the β4-sheet also seem to be important for Cdc37 binding (see Fig. 2).[27] Similar binding studies using truncated versions of Lck suggest that Cdc37 interacts with a region that included the α-C helix and the loop between this helix and the β4-sheet.[46] Hsp90 appears to interact with distinct portions of a misfolded kinase, including the linker between the N- and C-lobes (between β5 and helix D).[46,47] It is perhaps worth noting that neither Cdc37 nor Hsp90 interact with properly folded kinases[48] from which these structural designations are taken—and that the precise conformations recognized by the chaperones are unknown.

The effect of Cdc37 loss of function on kinase stability has been well addressed by two different approaches. In yeast, loss of Cdc37 function leads to rapid degradation of some client kinases,[7] but not all of them.[49] In mammalian systems, mutation of Ser13 to Ala also leads to kinase destabilization.[25] These data suggest that failure of Cdc37 to interact with client kinases leads to their rapid clearance from the cell. This effect is not restricted to loss of Cdc37 function, since inhibition of Hsp90 with geldanamycin has a similar effect on many but not all kinases.[10,50] Interestingly, geldanamycin does not cause dissociation of Hsp90 from Cdc37 itself, but it does stimulate removal of a client kinase from the complex. This suggests that both chaperones function together in the maturation of client kinases. On the other hand, results from genetic studies also suggest a model whereby Cdc37 might function independently of Hsp90. Foremost among these observations is that overexpression of Cdc37 suppresses defects arising from mutation in yeast Hsp90.[23,42] However, this effect works in this one direction only, and overexpressing Hsp90 does not suppress phenotypes in *cdc37* mutants.[42]

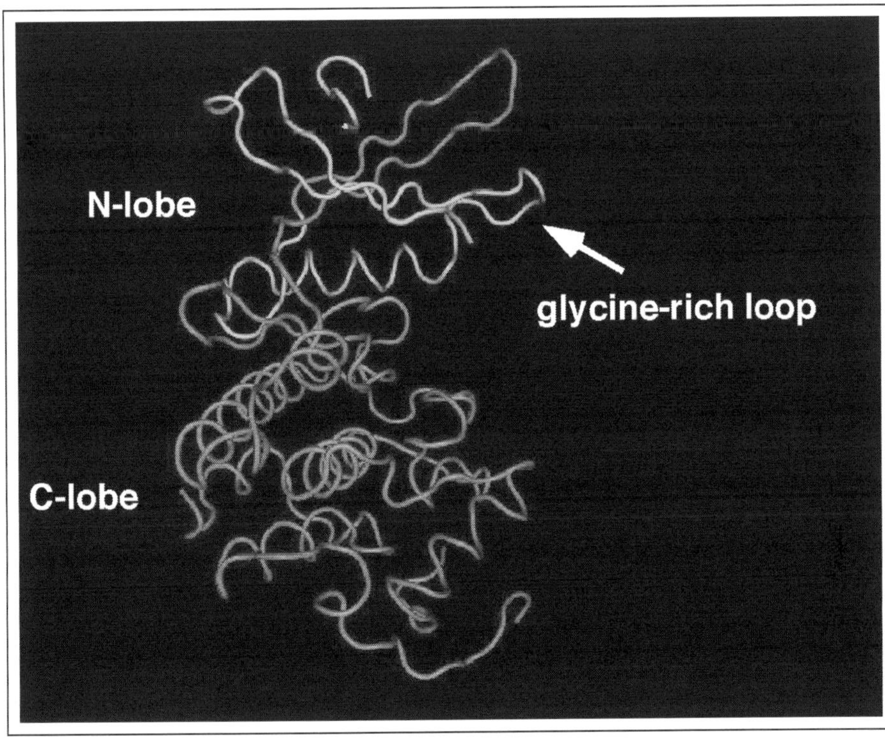

Figure 2. Structure of Cdk6 showing limits of Cdc37 binding as deduced from studies with Cdk4 and Lck. The parts of the kinase that interact with Cdc37 are shown in blue. The glycine-rich loop location is arrowed (adapted from refs. 43,46). Cdk6 structure is shown (PDB 1JOW) because there is no crystal structure of Cdk4, and Cdk6 is known to bind to Cdc37.[10] A color version of this figure is available online at http://www.eurekah.com.

The ability of Cdc37 to function independently of Hsp90 has been suggested by various genetic studies. For example, truncation of Cdc37 to a small N-terminal fragment of 148 amino acids was sufficient for cell viability (Fig. 3) and was capable of stabilizing v-Src levels in cells deleted for *STI1*[23] (also see ref. 22). These data suggest that Cdc37's chaperone activity is sufficient to stabilize a client kinase. However, the activity of the stabilized kinase was relatively low in cells expressing the Cdc37 truncation, although not as low compared to the *sti1Δ* cells alone. On the other hand, cells expressing a Cdc37 truncation that had the Hsp90 binding domain intact had both stabilized and active v-Src. This suggests that the N-terminal kinase binding domain of Cdc37 has a chaperone activity that can stabilize client kinases but cannot promote folding efficiently—for this Hsp90 is also required. However, since cells expressing $Cdc37^{1\text{-}148}$ are viable, it is clear that regulation of Hsp90's ATPase by the middle domain of Cdc37 is not essential for kinase maturation, although it may affect the efficiency of the folding reaction.

One further set of data suggests that Hsp90 and Cdc37 have a more complex relationship than simply inhibiting Hsp90's ATPase. Several studies have shown that expression of a truncated form of Cdc37 that is deleted for the Hsp90 binding domain ($Cdc37^{\Delta C}$; amino acids 1-163) results in a dominant negative effect on kinase maturation and cell growth.[13,48,51] This appears to result from failure of truncated Cdc37 to recruit Hsp90 to client kinases, although it can bind directly to the kinases themselves. However, one study has shown that $Cdc37^{\Delta C}$ can

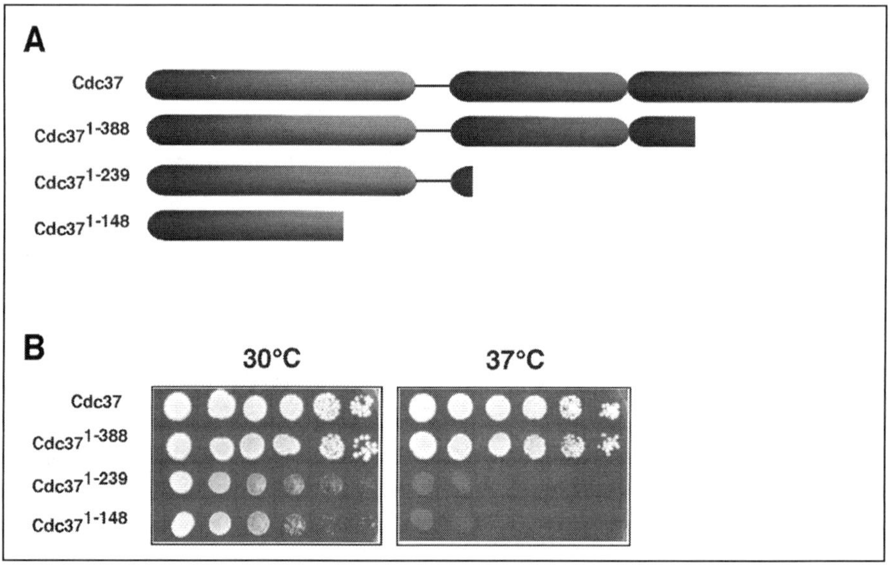

Figure 3. Domains of Cdc37 that are essential for viability in yeast. A) Schematic representation of truncations of yeast Cdc37. Domain organization as in Figure 1. B) Study showing viability of yeast grown under normal (30°C) and stressful (37°C) conditions. The images show colonies of yeast strains serially diluted on agar plates. Note that even the smallest truncations that are deleted for the Hsp90 binding domain mediate cell viability, although only under nonstress conditions. This suggests that the interaction of Cdc37 with Hsp90 is more important when cells are grown in a stressful environment. Reproduced with permission from: Lee P et al. J Cell Biol 2002; 159(6):1051-1059,[23] ©2002 The Rockefeller University Press. A color version of this figure is available online at http://www.eurekah.com.

actually stabilize a mutant form of Hck, a nonreceptor tyrosine kinase, in animal cells.[52] In this study, overexpression of Cdc37$^{\Delta C}$ also led to recruitment of Hsp90 to the mutant Hck. This finding is unexpected based on the notion that Cdc37$^{\Delta C}$ is deleted for the Hsp90 binding domain. There are two possible explanations for the ability of Cdc37$^{\Delta C}$ to stabilize the mutant Hck and recruit Hsp90. The first is that Cdc37$^{\Delta C}$ can still bind to Hsp90. Although the evidence for this is still weak, mutation of Trp7 of Cdc37 did affect complex formation with Hsp90.[20] The second possibility is that Cdc37 affects Hsp90 binding to client kinases via an allosteric mechanism. In this case, binding of Cdc37 to misfolded client kinases promotes a conformational change that reveals a binding site for Hsp90. The ability of Cdc37$^{\Delta C}$ to do this might be selective for different kinases. This is because the Cdc37$^{\Delta C}$ has a dominant negative phenotype for Raf and HRI, since it failed to promote the conformation with which Hsp90 could interact.[48,51] On the other hand, an even smaller truncation of Cdc37 (called D1; amino acids 1-126) bound to HRI in lysates but this binding was sensitive to geldanamycin—suggesting that Hsp90 was present in a ternary complex with the client.[20] It would appear, therefore, that Cdc37 and Hsp90 relate to each other via the client itself, or perhaps through a direct binding interaction that is distinct from the middle domain of Cdc37 interacting with the N-terminal ATP binding domain of Hsp90.

It is perhaps worth mentioning that several other chaperones and co-chaperones are needed for protein kinase folding/maturation beyond those mentioned above. Both biochemical and genetic studies point to the involvement of the Hsp70/Hsp40 chaperones, Hsp110, immunophilins and p23 in this process.[24,53-58] How these coordinate their efforts with Hsp90 awaits further exploration.

Acknowledgements

A.J. Caplan thanks Mary Kunjappu and Dr. Robert Matts for comments on the manuscript. Work in the authors' lab is supported by grants from the National Institutes of Health DK40698 and GM70596.

References

1. Hunter T, Poon RYC. Cdc37: A protein kinase chaperone? Trends Cell Biol 1997; 7:157-161.
2. MacLean M, Picard D. Cdc37 goes beyond Hsp90 and kinases. Cell Stress Chaperones 2003; 8(2):114-119.
3. Pearl LH. Hsp90 and Cdc37—A chaperone cancer conspiracy. Curr Opin Genet Dev 2005; 15(1):55-61.
4. Reed SI. The selection of amber mutations in genes required for completion of start, the controlling event of the cell division cycle of S. cerevisiae. Genetics 1980; 95(3):579-588.
5. Reed SI. The selection of S. cerevisiae mutants defective in the start event of cell division. Genetics 1980; 95(3):561-577.
6. Gerber MR, Farrell A, Deshaies RJ et al. Cdc37 is required for association of the protein kinase Cdc28 with G1 and mitotic cyclins. Proc Natl Acad Sci USA 1995; 92(10):4651-4655.
7. Farrell A, Morgan DO. Cdc37 promotes the stability of protein kinases Cdc28 and Cak1. Mol Cell Biol 2000; 20(3):749-754.
8. Whitelaw ML, Hutchison K, Perdew GH. A 50-kDa cytosolic protein complexed with the 90-kDa heat shock protein (hsp90) is the same protein complexed with pp60v-src hsp90 in cells transformed by the Rous sarcoma virus. J Biol Chem 1991; 266(25):16436-16440.
9. Brugge JS. Interaction of the Rous sarcoma virus protein pp60src with the cellular proteins pp50 and pp90. Curr Top Microbiol Immunol 1986; 123:1-22.
10. Stepanova L, Leng X, Parker SB et al. Mammalian p50Cdc37 is a protein kinase-targeting subunit of Hsp90 that binds and stabilizes Cdk4. Genes Dev 1996; 10(12):1491-1502.
11. Perdew GH, Wiegand H, Vanden Heuvel JP et al. A 50 kilodalton protein associated with raf and pp60(v-src) protein kinases is a mammalian homolog of the cell cycle control protein cdc37. Biochemistry 1997; 36(12):3600-3607.
12. Stepanova L, Yang G, DeMayo F et al. Induction of human Cdc37 in prostate cancer correlates with the ability of targeted Cdc37 expression to promote prostatic hyperplasia [In Process Citation]. Oncogene 2000; 19(18):2186-2193.
13. Schwarze SR, Fu VX, Jarrard DF. Cdc37 enhances proliferation and is necessary for normal human prostate epithelial cell survival. Cancer Res 2003; 63(15):4614-4619.
14. Rao J, Lee P, Benzeno S et al. Functional interaction of human Cdc37 with the androgen receptor but not with the glucocorticoid receptor. J Biol Chem 2001; 276(8):5814-5820.
15. Wang X, Grammatikakis N, Hu J. Role of p50/CDC37 in hepadnavirus assembly and replication. J Biol Chem 2002; 277(27):24361-24367.
16. Yun BG, Matts RL. Differential effects of Hsp90 inhibition on protein kinases regulating signal transduction pathways required for myoblast differentiation. Exp Cell Res 2005; 307(1):212-223.
17. Caplan AJ. What is a co-chaperone? Cell Stress Chaperones 2003; 8(2):105-107.
18. Roe SM, Ali MM, Meyer P et al. The Mechanism of Hsp90 regulation by the protein kinase-specific co-chaperone p50(cdc37). Cell 2004; 116(1):87-98.
19. Zhang W, Hirshberg M, McLaughlin SH et al. Biochemical and structural studies of the interaction of Cdc37 with Hsp90. J Mol Biol 2004; 340(4):891-907.
20. Shao J, Irwin A, Hartson SD et al. Functional dissection of cdc37: Characterization of domain structure and amino acid residues critical for protein kinase binding. Biochemistry 2003; 42(43):12577-12588.
21. Roiniotis J, Masendycz P, Ho S et al. Domain-mediated dimerization of the Hsp90 co-chaperones Harc and Cdc37. Biochemistry 2005; 44(17):6662-6669.
22. Turnbull EL, Martin IV, Fantes PA. Cdc37 maintains cellular viability in Schizosaccharomyces pombe independently of interactions with heat-shock protein 90. Febs J 2005; 272(16):4129-4140.
23. Lee P, Rao J, Fliss A et al. The Cdc37 protein kinase-binding domain is sufficient for protein kinase activity and cell viability. J Cell Biol 2002; 159(6):1051-1059.
24. Bandhakavi S, McCann RO, Hanna DE et al. A positive feedback loop between protein kinase CKII and Cdc37 promotes the activity of multiple protein kinases. J Biol Chem 2003; 278(5):2829-2836.
25. Miyata Y, Nishida E. CK2 controls multiple protein kinases by phosphorylating a kinase-targeting molecular chaperone, Cdc37. Mol Cell Biol 2004; 24(9):4065-4074.

26. Shao J, Prince T, Hartson SD et al. Phosphorylation of serine 13 is required for the proper function of the Hsp90 co-chaperone, Cdc37. J Biol Chem 2003; 278(40):38117-38120.
27. Zhao Q, Boschelli F, Caplan AJ et al. Identification of a conserved sequence motif that promotes Cdc37 and Cyclin D1 binding to Cdk4. J Biol Chem 2004; 279(13):12560-12564.
28. Brugge JS, Jarosik G, Andersen J et al. Expression of Rous sarcoma virus transforming protein pp60v-src in Saccharomyces cerevisiae cells. Mol Cell Biol 1987; 7:2180-2187.
29. Siligardi G, Panaretou B, Meyer P et al. Regulation of Hsp90 ATPase activity by the co-chaperone Cdc37p/p50cdc37. J Biol Chem 2002; 277(23):20151-20159.
30. Prodromou C, Pearl LH. Structure and functional relationships of Hsp90. Curr Cancer Drug Targets 2003; 3(5):301-323.
31. Pratt WB, Toft DO. Regulation of signaling protein function and trafficking by the hsp90/hsp70-based chaperone machinery. Exp Biol Med (Maywood) 2003; 228(2):111-133.
32. Hartson SD, Irwin AD, Shao J et al. p50(c)(dc37) Is a nonexclusive Hsp90 cohort which participates intimately in Hsp90-mediated folding of immature kinase molecules. Biochemistry 2000; 39(25):7631-7644.
33. Panaretou B, Siligardi G, Meyer P et al. Activation of the ATPase activity of hsp90 by the stress-regulated co-chaperone aha1. Mol Cell 2002; 10(6):1307-1318.
34. Lotz GP, Lin H, Harst A et al. Aha1 binds to the middle domain of hsp90, contributes to client protein activation, and stimulates the ATPase activity of the molecular chaperone. J Biol Chem 2003; 278(19):17228-17235.
35. Harst A, Lin H, Obermann WM. Aha1 competes with Hop, p50 and p23 for binding to the molecular chaperone Hsp90 and contributes to kinase and hormone receptor activation. Biochem J 2005; 387(3):789-796.
36. Siligardi G, Hu B, Panaretou B et al. Co-chaperone regulation of conformational switching in the Hsp90 ATPase cycle. J Biol Chem 2004; 279(50):51989-51998.
37. Prodromou C, Siligardi G, O'Brien R et al. Regulation of Hsp90 ATPase activity by tetratricopeptide repeat (TPR)-domain co-chaperones. EMBO J 1999; 18(3):754-762.
38. Abbas-Terki T, Briand PA, Donze O et al. The Hsp90 co-chaperones Cdc37 and Sti1 interact physically and genetically. Biol Chem 2002; 383(9):1335-1342.
39. Lee P, Shabbir A, Cardozo C et al. Sti1 and Cdc37 can stabilize Hsp90 in chaperone complexes with a protein kinase. Mol Biol Cell 2004; 15:1785-1792.
40. Nicolet CM, Craig EA. Isolation and characterization of STI1, a stress-inducible gene from Saccharomyces cerevisiae. Mol Cell Biol 1989; 9(9):3638-3646.
41. Felts SJ, Toft DO. p23, a simple protein with complex activities. Cell Stress Chaperones 2003; 8(2):108-113.
42. Kimura Y, Rutherford SL, Miyata Y et al. Cdc37 is a molecular chaperone with specific functions in signal transduction. Genes Dev 1997; 11(14):1775-1785.
43. Mort-Bontemps-Soret M, Facca C, Faye G. Physical interaction of Cdc28 with Cdc37 in Saccharomyces cerevisiae. Mol Genet Genomics 2002; 267(4):447-458.
44. Silverstein AM, Grammatikakis N, Cochran BH et al. p50(cdc37) binds directly to the catalytic domain of Raf as well as to a site on hsp90 that is topologically adjacent to the tetratricopeptide repeat binding site. J Biol Chem 1998; 273(32):20090-20095.
45. Akamine P, Madhusudan, Wu J et al. Dynamic features of cAMP-dependent protein kinase revealed by apoenzyme crystal structure. J Mol Biol 2003; 327(1):159-171.
46. Prince T, Matts RL. Definition of protein kinase sequence motifs that trigger high affinity binding of Hsp90 and Cdc37. J Biol Chem 2004; 279(38):39975-39981.
47. Scroggins BT, Prince T, Shao J et al. High affinity binding of Hsp90 is triggered by multiple discrete segments of its kinase clients. Biochemistry 2003; 42(43):12550-12561.
48. Shao J, Grammatikakis N, Scroggins BT et al. Hsp90 regulates p50(cdc37) function during the biogenesis of the active conformation of the heme-regulated eIF2 alpha kinase. J Biol Chem 2001; 276(1):206-214.
49. Schutz AR, Giddings Jr TH, Steiner E et al. The yeast CDC37 gene interacts with MPS1 and is required for proper execution of spindle pole body duplication. J Cell Biol 1997; 136(5):969-982.
50. Tikhomirov O, Carpenter G. Identification of ErbB-2 kinase domain motifs required for geldanamycin-induced degradation. Cancer Res 2003; 63(1):39-43.
51. Grammatikakis N, Lin JH, Grammatikakis A et al. p50(cdc37) acting in concert with Hsp90 is required for Raf-1 function. Mol Cell Biol 1999; 19(3):1661-1672.
52. Scholz G, Hartson SD, Cartledge K et al. p50(Cdc37) can buffer the temperaturesensitive properties of a mutant of Hck. Mol Cell Biol 2000; 20(18):6984-6995.

53. Nair SC, Toran EJ, Rimerman RA et al. A pathway of multi-chaperone interactions common to diverse regulatory proteins: Estrogen receptor, Fes tyrosine kinase, heat shock transcription factor Hsf1, and the aryl hydrocarbon receptor. Cell Stress Chaperones 1996; 1(4):237-250.
54. Thulasiraman V, Yun BG, Uma S et al. Differential inhibition of Hsc70 activities by two Hsc70-binding peptides. Biochemistry 2002; 41(11):3742-3753.
55. Donze O, Picard D. Hsp90 binds and regulates Gcn2, the ligand-inducible kinase of the alpha subunit of eukaryotic translation initiation factor 2 [corrected]. Mol Cell Biol 1999; 19(12):8422-8432.
56. Donze O, Abbas-Terki T, Picard D. The Hsp90 chaperone complex is both a facilitator and a repressor of the dsRNA-dependent kinase PKR. EMBO J 2001; 20(14):3771-3780.
57. Dey B, Caplan AJ, Boschelli F. The Ydj1 molecular chaperone facilitates formation of active p60v-src in yeast. Mol Biol Cell 1996; 7(1):91-100.
58. Kimura Y, Yahara I, Lindquist S. Role of the protein chaperone YDJ1 in establishing Hsp90-mediated signal transduction pathways. Science 1995; 268(5215):1362-1365.

Chapter 6

UNC-45:
A Chaperone for Myosin and a Co-Chaperone for Hsp90

Odutayo O. Odunuga and Henry F. Epstein*

Abstract

UNC-45 is a prototype of the protein family characterized by the presence of the C-terminal UCS (UNC-45/CRO1/She4p) domain. These proteins function in various important actin- and myosin-dependent cellular processes that include myofibril organization and muscle functions, cell differentiation, embryonic development, cytokinesis and endocytosis. Mutations in the genes that code for UCS domain proteins cause serious defects in these actomyosin-based processes. Homologs of UCS domain proteins have been identified in fungi, nematodes, insects, fish, amphibians, birds and mammals. In addition to the UCS domain, the animal homologs (UNC-45) contain an N-terminal TPR domain and a conserved central region. UNC-45 has been shown to act as chaperone to fold the heads of myosin heavy chain of various types. Apart from assisting myosin heads to fold correctly, UNC-45 is known to bind Hsp90 directly and several UCS protein complexes appear to be dependent on the Hsp90 chaperone machinery. These findings suggest that UNC-45 and other proteins containing the UCS domain are a new class of Hsp90 co-chaperones.

Introduction

UNC-45 is a prototype of a class of proteins known as the UCS- (UNC-45 in *Caenorhabditis elegans*, CRO1 in *Podospora anserina* and She4p in *Saccharomyces cerevisiae*) domain containing proteins.[1,2] The UCS-domain containing proteins are emerging as essential for a wide spectrum of myosin- and actin-related cellular processes in many eukaryotes, ranging from fungi to humans (Table 1). They are necessary for important cellular processes such as myofibril organization, cell differentiation, embryonic development, cytokinesis, endocytosis, and syncytial-cellular stage transition.[1,2] The UCS proteins can be divided into two broad sub-classes; animal and fungal UCS-containing proteins. The only similarity between these two sub-classes of proteins is the presence of the homologous C-terminal UCS domain. The animal UCS proteins contain an N-terminal tetratricopeptide repeat (TPR)[3] domain which is absent in the fungal proteins. While, only one copy of the gene is found in invertebrates, vertebrates have two copies encoding differentially expressed isoforms.[4] Mutations in the UCS proteins result in various defective actomyosin-based processes such as cytokinesis in *Schizosaccharomyces pombe*,[5] endocytosis and trafficking in *S. cerevisiae*,[6,7] syncytial-cellular stage transition in *P. anserina*,[8] and myofibril organization and cytokinesis in *C. elegans*.[9,10] *C. elegans* UNC-45 has been established as a chaperone for the motor domain of myosin.[11] Other UCS proteins such as Rng3p in

*Corresponding Author: Henry F. Epstein—Department of Neuroscience and Cell Biology, University of Texas Medical Branch at Galveston, 301 University Boulevard, Galveston, Texas 77555-0625, U.S.A. Email: hepstein@utmb.edu

Networking of Chaperones by Co-Chaperones, edited by Gregory L. Blatch.

Table 1. UCS domain containing proteins

UCS Proteins	Organisms	Type of Interacting Myosin	Loss of Function Phenotypes	Refs.
General cell (GC) UNC-45	Vertebrates (Mouse, Humans)	Cytoskeletal myosin II	Inhibition of cell proliferation and fusion	4
Striated muscle (SM) UNC-45	Vertebrates (Mouse, Humans)	Sarcomeric myosin II	Loss of sarcomere organization	4
DUNC-45	*D. melanogaster* (Fruitfly)	Sarcomeric myosin II	Embryonic and late larval stage lethality, reduced body size and defects in mobility	14
UNC-45	*C. elegans* (Soil Nematode)	Cytoskeletal and sarcomeric myosin II	Fewer thick filaments, myofibril disorganization, paralysis and failure of embryonic cytokinesis	9,41,42, 49,50
Rng3p	*S. pombe* (Fission yeast)	Cytoskeletal myosin II	Defective actomyosin ring, failure of cytokinesis	5,12,36, 54,57
She4p	*S. cerevisiae* (Budding yeast)	Myosin types I & V	Loss of actin polarization of cell, defective internalization of membrane, defects in endocytosis	6,7,13, 56
CRO1	*P. anserina* (Filamentous fungus)	?	Inability to form septum, defective syncytial-cellular transition	8

S. pombe,[12] She4p in *S. cerevisiae*,[13] and DUNC-45[14] in *Drosophila melanogaster* have also been shown to interact with myosin and modulate its functions. The interaction of UCS proteins is not limited to sarcomeric myosins alone; cytoskeletal myosins including both conventional (filament assembling) myosin II and unconventional (nonassembling) myosins I and V, are known to require UCS proteins for their proper functions (Fig. 1A).[10,13] Myosins may require the UCS-containing chaperones due to their large size, complexity of their structures and the need to form highly organized oligomeric assemblies which are sometimes composed of different isoforms and other associating proteins.[15] Furthermore, UCS proteins may recruit Hsp90 to form a ternary complex with myosin,[11,12] although the physiological role of this association is not understood. In this chapter, we discuss the functions of UNC-45 proteins as chaperones for myosin and co-chaperones in targeted chaperone systems, genetic and biochemical studies carried out on UCS proteins in various organisms, and their involvement in cell differentiation, embryonic development, muscle functions and other actomyosin-dependent processes.

UNC-45 and Myosin Folding, Assembly and Function

The myosin family is a large group of motor proteins that interact with actin, hydrolyze ATP and produce movement along the actin filament. Myosins are involved in a broad spectrum of cellular processes that include cellular trafficking, phagocytosis, muscle contraction, cytokinesis and cytoskeletal assembly. The full protein complement of a myosin is composed of two parts: the myosin heavy chains (MHC) and the myosin light chains (MLC). Typically, a myosin heavy chain is comprised of three functional regions: (1) a conserved (catalytic) motor or head that contains actin and ATP binding sites, (2) the neck domain which binds myosin light chains, and (3) the tail domain, an α-helical coiled-coil rod that serves to anchor and position the motor domain to interact with actin (Fig. 1B). Myosin II includes the classical conventional myosin first isolated from muscle, but subsequently found in nonmuscle

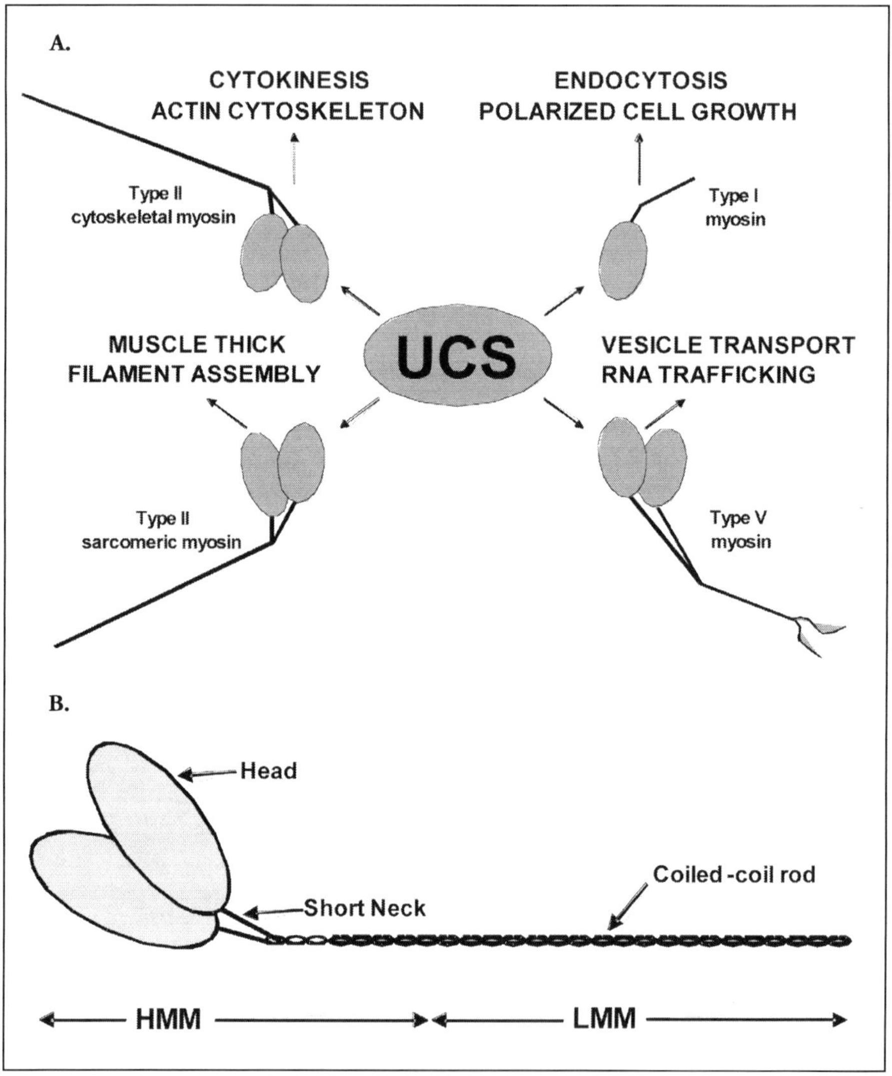

Figure 1. A) UCS domain-myosin interactions. UCS domains from *C. elegans* (UNC-45), *S. pombe* (Rng3p) and *S. cerevisiae* (She4p) interact with various types of myosin. B) Domain organization in type Myosin II. The head and the neck alone are sufficient for motor activity. Myosin light chains interact with the short neck. Other myosin types e.g., Myosin V have longer necks. HMM and LMM represent heavy meromyosin and light meromyosin, respectively.

cells and protists.[16] The sarcomeric muscle myosin II is the only member of the myosin family that is assembled into the thick filaments of skeletal and cardiac muscles. Myosin is a multidomain protein; therefore its folding pathway may be expected to be complex.[17] The myosin head itself contains multiple domains connected by flexible loops[18,19] while the light chains and rod are simpler in structure. Myosin light chains and rod when expressed in bacteria, fold into functionally active structures.[20,21] Regardless of their origin, expression of myosin motors has proved difficult in bacteria; this may be due to the complex nature of their structure.[22,23] Using the baculovirus expression system, in insect cells, considerable success has been achieved in expressing heavy meromyosin (HMM; Fig. 1B) from cytoskeletal types II, V and VI, and to a limited extent, cardiac sarcomeric myosins.[24-26] These HMMs have properly folded motor domains that are capable of binding actin.[24-26] However, the fast skeletal muscle myosin head has been expressed in active form in C2C12 mouse myogenic cell line suggesting that this myosin requires additional cofactor(s) which are present in myogenic cells for folding.[27,28] In vitro, chimeric fast skeletal muscle myosin head fused to green fluorescent protein (GFP) folds very slowly and transits through multiple intermediates in a temperature-dependent manner that strongly suggests a high susceptibility to off-pathway interactions and aggregations and hence the need for chaperone-assisted folding.[28] Expression of the protein in vivo is cell-dependent: C2C12 myogenic cell lines yield a folded and active protein that exhibits Mg^{2+} ATP-sensitive actin-binding and myosin motor activity, while epithelial cell lines yield inactive protein aggregates.[28] This observation suggests that the myosin motor requires cytosolic molecular chaperones to fold correctly under physiological conditions and that the required factor(s) are optimized in muscle cells.[28] In addition, during de novo folding and assembly of striated muscle myosin heavy chain, Hsp70 and Hsp90 colocalize with the myosin intermediates but not the mature myofibrils, though this does not implicate direct physical interaction between either chaperone and myosin.[29]

UNC-45 and the Molecular Chaperone Hsp90

Biochemical and genetic evidence confirm that UNC-45 and other UCS containing proteins interact with Hsp90 chaperone.[11,12,29] Full-length UNC-45 from *C. elegans* binds both endogenous Hsp70 and Hsp90 from Sf9 insect cell lysates.[11] Mutant UNC-45 protein lacking the TPR domain interacts with Hsp70 but not Hsp90 also from Sf9 insect cell lysates, indicating that the interaction of UNC-45 with Hsp90 requires the TPR domain and that the Hsp70-binding site lies within the central/UCS region.[11] In surface plasmon resonance spectroscopy experiments, the binding of Hsp90 to the TPR domain of UNC-45 is preferentially competed by Hsp90 C-terminal peptides in comparison to the analogous Hsp70 peptides.[11] At 30°C, purified Hsp90, myosin and UNC-45 can form the three possible binary complexes in pull-down assays.[11] UNC-45 possesses chaperone activity which maps to the central/UCS region of the protein, and demonstrates prevention of thermal aggregation of both myosin head and citrate synthase (Fig. 2A).[11] Therefore both Hsp90 and UNC-45 are capable of interacting directly with the myosin head and exerting chaperoning activities. The myosin-binding site of UNC-45 also lies within the central/UCS regions. The interaction of UNC-45 with Hsp70 and Hsp90 may occur via two mechanisms. First, it may be a classical chaperone-client interaction, especially with the promiscuous Hsp70 which binds nonspecifically to several recombinantly expressed proteins to fold them. Second, the interaction may be specific co-chaperone-chaperone association such as the TPR-based interaction between the Hsp70/Hsp90 organizing protein (Hop) and Hsp70 and Hsp90.[30] The specific binding of UNC-45 to Hsp90 appears to be a direct co-chaperone-chaperone interaction between these proteins, respectively. On the other hand, Hsp70 may be acting simply as a chaperone to fold the recombinant UNC-45 protein. We propose that similarly to Hop[30,31] (see Chapter by Daniel et al) UNC-45 acts as an adaptor to bring Hsp90 in close proximity to the myosin protein (Fig. 2B). Many oligomeric protein assemblies are mediated by Hsp90 chaperone and associated protein partners.[32] The ability of UNC-45 to perform this role depends on its possession of a TPR domain, a structural domain mediating protein-protein interactions and found in several proteins that interact with Hsp90.[33] However, unlike Hop which dissociates from the steroid receptor in the presence of Hsp90 and other steroid-associating proteins,[31,34] UNC-45 seems to perform more than a targeting role; its chaperoning activity on myosin heads may be necessary for myosin assembly, contractile function and turnover in vivo.[35] Recent studies on *S. pombe* show that recombinant Rng3p protein activates the contractile function of myosin in vitro;[36] this observation is consistent with our proposal. Also, in a yeast two hybrid assay, *S. cerevisiae* She4p was found to interact with Hsp90.[37] The ability of UNC-45 to chaperone myosin heads makes it similar to several proteins such as p23,[38]

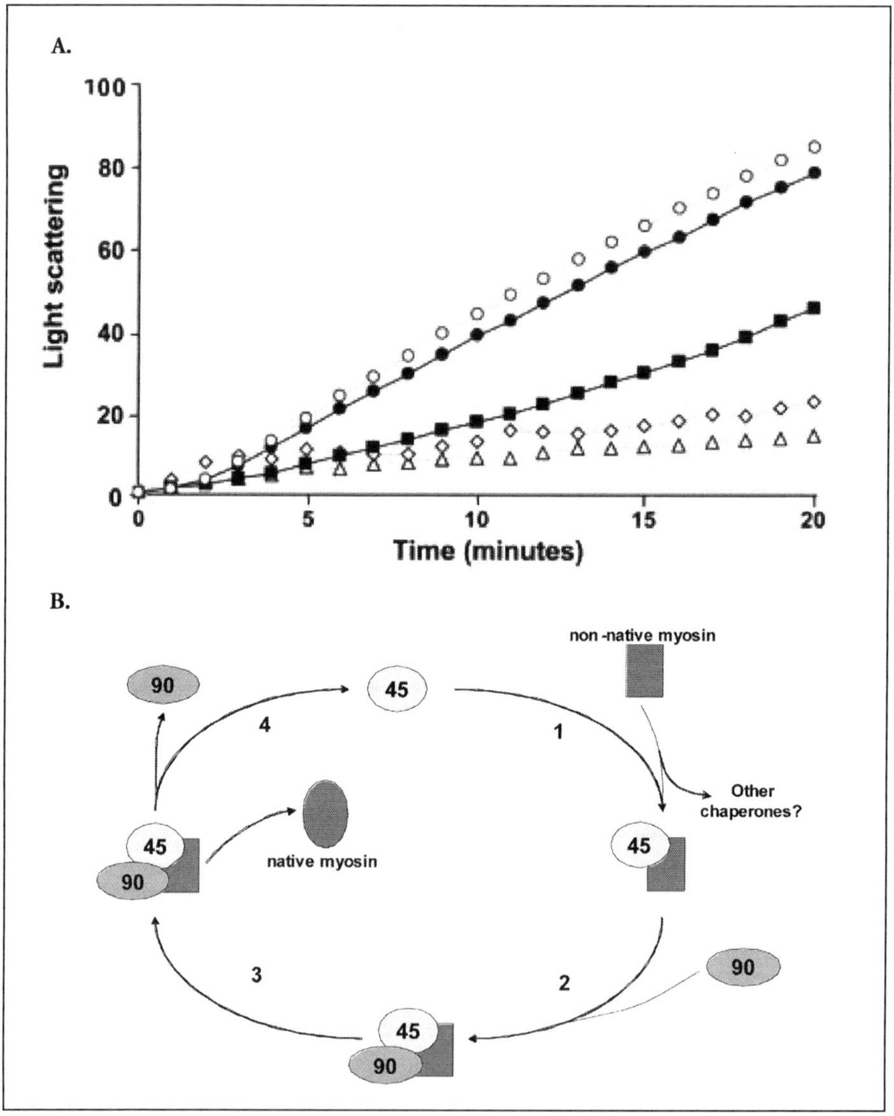

Figure 2. A) UNC-45 exhibits chaperone activity for myosin head. UNC-45 prevents aggregation of myosin head (1.0 μM) measured by light scattering (320 nm) with no additional protein (solid circles), 2.0 μM bovine serum albumin (open circles), 0.5 μM UNC-45 (squares), 1.0 μM UNC-45 (diamonds) and 2.0 μM UNC-45 (triangles). Experiments were carried out at 43°C.[11] B) Proposed model for the function of UNC-45 in the folding and assembly of myosin motors. The model is based on the mechanism of maturation of steroid hormone receptors.[58] Nascent or unfolded myosin may interact first with other chaperones or chaperonins (1). UNC-45 targets Hsp90 (2) chaperone machinery to partially unfolded myosin, and may also exert its own chaperone activity on the myosin molecule. Hsp90 and UNC-45 may promote further folding and stability of the near-native myosin molecule (3). After the myosin reaches a fully native state, both Hsp90 and UNC-45 may dissociate from the complex to begin another cycle (4). In addition, UNC-45 may also associate with assembled thick filaments, possibly related to the dynamics of myosin assembly, disassembly, and turnover.

immunophilins[39] and Cdc37[40] which are capable of binding directly to Hsp90 as well as exerting chaperone effects on their clients.

UNC-45 Proteins in Invertebrates: *C. elegans* and *D. melanogaster*

The *unc-45* gene was originally identified in *C. elegans* through the recessive, temperature-sensitive (*ts*) mutant allele, *e286*.[9] *C. elegans* possesses only one copy of the *unc-45* gene. The E286 mutant worms are paralyzed, with disorganized thick filaments in their muscles when grown at 25°C, but at the permissive temperature of 15°C, the worms display phenotypes essentially similar to the wild-type.[9] These phenotypes can be reversed by temperature shifts in developing embryos and larvae but not in adult worms, implying that UNC-45 possesses a function essential for proper myofilament arrays to form.[9] Detailed genetic analysis revealed three additional recessive (*ts*) mutations and two lethal mutations in the *unc-45* gene.[41,42] All of the (*ts*) alleles, which show similar effects on myofibril formation, contain missense substitutions in the C-terminal region of the UNC-45 protein.[42] The lethal alleles each contain a stop codon located within the central region of the protein (Fig. 3) preventing further translation of the *unc-45* gene product.[42] Genetic analysis suggests functional relationships between the protein products of *unc-45* and of the *unc-54* and *myo-3* genes, which code for myosin heavy chains A and B respectively, that form homodimeric myosins in the body wall muscle of *C. elegans*.[43] The *unc-45* (*ts*) mutants may directly affect myosin B, the major isoform, by generating an incorrectly folded myosin B which drastically reduces the number of intact thick filaments and therefore incapable of forming proper myofilament assemblies.[43,44] Null mutations in the *unc-54* gene generate defects in muscle structure and functions similar to that of the *unc-45* (*ts*) alleles, implying that the two genes may be epistatic.[43] In the normal *C. elegans* thick filament, the two myosin isoforms are differentially assembled such that myosin B flanks a central myosin A zone.[45,46] However, in worms harboring the *unc-45* (*ts*) mutant genes, this differential assembly is lost and instead, there is a scrambling of the myosins,[42] which might be due to improper folding of the myosins and consequent decreases in their concentrations. The lethal *unc-45* alleles cause arrest of development at the two-fold embryonic stage resulting in inability to produce functional body wall muscle.[41]

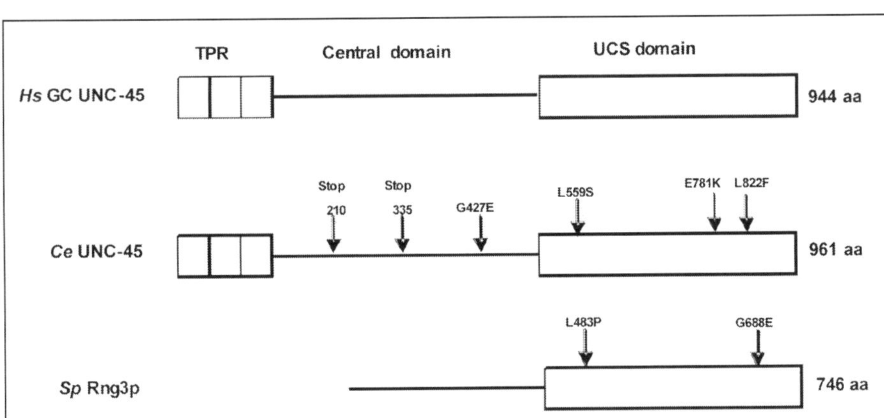

Figure 3. Structure and domain organization of UNC-45 homologs. *Hs* GC UNC-45, *Ce* UNC-45 and *Sp* Rng3p represent the human general cell UNC-45 (NP_061141),[4] *C. elegans* UNC-45 (NP_497205)[9,11] and *S. pombe* ring assembly protein 3 (O74994),[5] respectively. The TPR and UCS domains are represented by small and large boxes, respectively. The horizontal line represents the central domain. The positions of amino acid substitutions are indicated by vertical arrows.

Interestingly, mutant worms lacking the essential *myo-3* encoded myosin heavy chain A, the minor isoform of myosins found in *C. elegans* body wall muscle, also manifest severely impaired thick filament assembly with little or no body movement.[47] This observation suggests the necessity for myosin A in the formation of the nematode thick filament; moreover, myosin B cannot substitute for myosin A to reverse the phenotype.[47] In addition, decreased pharyngeal pumping in worms containing the *unc-45* (*ts*) mutant genes suggests that myosins C and D, which are exclusively found in the pharyngeal muscles of *C. elegans*,[48] may be affected.[41] Localization by both antibodies and GFP tagging show that UNC-45 protein is expressed in all *C. elegans* muscle cells at the adult stage.[49,50] In the developing body wall muscle of *C. elegans* larvae, UNC-45 is found in the cytosol, whereas in the mature adult muscle, it is localized to the A bands of the sarcomere, apparently chiefly with myosin B.[50] *C. elegans* UNC-45 also colocalizes in the cleavage furrow with the conventional cytoskeletal myosin II, a protein that plays an essential role during embryonic cytokinesis.[51] RNA interference studies reveal that the UNC-45 protein (1) is maternally contributed, hence rescue can occur to some extent and, (2) that it plays a role in cytokinesis in addition to muscle development.[10]

In *D. melanogaster* embryos, high levels of *dunc-45* (*D. melanogaster* homolog of *unc-45* gene) RNA are present in mesodermal precursors to muscle; with accumulation in other tissues as well.[14] This suggests that *dunc-45* gene product (DUNC-45) may be important in multiple cell types. Similarly to *C. elegans*, *D. melanogaster* possesses only one copy of the *dunc-45* gene. DUNC-45 is constitutively expressed during development in *D. melanogaster* and peaks at pupation, when adult tissues are being formed. Mutations in the *dunc-45* gene cause embryonic as well as late larval stage lethality. Reduced body size and defects in motility appear to be the results of embryonic body wall muscle dysfunction.[14]

UNC-45 Proteins in Vertebrates: Mouse and Human

Vertebrates have two copies of *unc-45*-like genes encoding distinct isoforms of UNC-45.[4] The genes are designated as *UNC45A* and *UNC45B* in human, and *Unc45a* and *Unc45b* in mouse respectively. In mouse, *Unc45a* encodes an isoform that is expressed in multiple adult organs including uterus, kidney, lung and liver, hence the designation general cell (GC) UNC-45.[4] The second isoform encoded by *Unc45b* is found almost exclusively in heart and skeletal muscles, and was therefore designated as striated muscle (SM) UNC-45.[4] The two isoforms share 50-55% sequence identity within both human and mouse. There is >90% sequence identity among similar isoforms between these species. When compared with *C. elegans* UNC-45, both isoforms show 30-40% identity with the worm protein. In the eight-day old mouse embryo, SM UNC-45 is predominantly expressed in the contractile heart and is hardly found in other organs; whereas GC UNC-45 is diffusely expressed and later concentrates in regions of intense development such as the branchial arches and the forelimb bud.[4] In C2C12 myogenic cells, only GC UNC-45 mRNA is detected in proliferating myoblasts, with the level decreasing as the cells progress to form myotubes.[4] In contrast, SM UNC-45 mRNA is detected only after the cells start fusing, peaking in young myotubes and dropping off as the myotubes age.[4] These observations in both mouse embryos and C2C12 myoblasts implicate stage-specific expression and functions of the UNC-45 isoforms in embryogenesis and muscle differentiation. The GC UNC-45 isoform may be involved in cell division and related cytoskeletal functions while the SM isoform may be related to striated muscle differentiation and myofibril formation. In fact, in C2C12 knock-down experiments using anti-sense oligonucleotides, GC UNC-45 antisense severely reduces myoblast proliferation and fusion while SM UNC-45 antisense results in significant loss of sarcomere organization.[4] Interestingly, reduction of SM UNC-45 mRNA did not affect the level of skeletal myosin heavy chain, whereas lowering of GC UNC-45 levels by antisense did.[4] Low levels of GC UNC-45 result in reduction of cell proliferation and differentiation which decreases the expression of sarcomeric myosin. The expression of SM UNC-45, however, starts at the fusion stage; therefore it may not affect myosin synthesis but rather its organization into thick filaments. Thus the expression of the two UNC-45 isoforms separates myosin synthesis from its organization into myofilaments.

On-going experiments are testing whether the two mammalian isoforms function independent of one another or whether there is any redundancy in their functions. The functions of the two isoforms may not necessarily be exclusive of one another, because the cytoskeleton is needed for myofibril formation and maintenance.

Structure of UNC-45 Proteins

The UNC-45 protein is predicted to contain three regions: a distinct N-terminal domain characterized by the presence of three TPR repeats (Fig. 4A), a central region, and a C-terminal UCS domain.[42] The TPR repeat is a degenerate motif that occurs in multiple copies in proteins and forms scaffolds that mediate protein-protein interactions.[33,52,53] The TPR region of UNC-45 preferentially binds Hsp90.[11] Both the central and the UCS domains of *C. elegans* UNC-45 share strong sequence conservation only with other animal UCS-containing proteins. The UCS domain was identified based on the positions of (*ts*) mutations in *C. elegans* UNC-45 and the presence of blocks of sequence identity between UNC-45, CROI and She4p (Fig. 4B).[42,54] It was confirmed by the presence of analogous (*ts*) mutations in *S. pombe* Rng3p.[54] The two UNC-45 null mutations that result in stop codons are both located in the central domain while three of the four UNC-45 (*ts*) mutations are found in the UCS domain. *C. elegans* and other animal UNC-45 proteins share conservation with fungal UCS-proteins only at key sites within the UCS domains. Hence it seems plausible that the myosin-chaperoning activity and involvement of the UCS-domain proteins in cytokinesis is mediated via their UCS domains. The central domain may act as a regulator of the activity of the protein, possibly through phosphorylation or other forms of protein modification. In addition, this domain could act to connect or transduce Hsp90 action to targeting activity.

Fungal UCS-Containing Proteins

UCS domain proteins have been identified in *S. cerevisiae* (She4p),[6,7] *P. anserina* (CRO1),[8] and *S. pombe* (Rng3p).[5] Sequence similarity among the three fungal proteins and UNC-45 is restricted to the C-terminal UCS domain. A striking difference between UNC-45 and the fungal UCS proteins is the absence of TPR domains in the latter. Despite the lack of TPR motifs, the N-terminal sequences of fungal UCS proteins may contain sequences capable of recruiting molecular chaperones.[55] However, all four UCS proteins are linked by their common association with cellular processes involving myosins. Although sequence similarity among the fungal UCS proteins is low,[36] the (*ts*) mutations in *S. pombe*, like *C. elegans*, are in conserved residues (Fig. 4B).

The *she4p* gene was identified and named differently in two independent screens in *S. cerevisiae*. The first screen was for the expression of the HO endonuclease in mother cells yielding the She4p-encoding gene (*SHE*; Swi5p-dependent HO expression),[6] and the other for defects in endocytosis identifying the *dim1* gene (*dim*; defective internalization of membrane).[7] In each case, only one copy of the mutant allele was found. Both null and (*ts*) mutants of the *she4p* gene cause defects in endocytosis and loss of actin polarization in the cell. Two-hybrid and biochemical experiments showed that She4p interacts, via its UCS domain, with the motor domains of conventional type II myosin (Myo1p) as well as unconventional types 1 (Myo3p/Myo5p) and V (Myo2p/Myo4p) myosins in an actin-dependent manner for proper endocytosis and cytokinesis to occur.[13,56] This observation suggests further that UCS proteins in general interact with myosin through their C-terminal regions. In addition, She4p interacts with Hsp90 in yeast two-hybrid assays.[37] The She4p protein is composed of 789 amino acids and shares about 33% similarities with other fungal UCS proteins.[36]

The *rng3* gene was identified in a large-scale screen for genes whose products function in cytokinesis.[5] The gene encodes a protein of 746 amino acid residues. Actin ring formation was found defective in *S. pombe* cells harboring mutants of both *rng3* (*rng3-65*) and *rng5* (*rng5-E1*) which codes for myo2, suggesting a functional interaction between the protein products of the two genes. Null mutants in *rng3* resemble deletion mutants in *myo2* (a type II myosin heavy chain) while (*ts*) *rng3* mutants show strong adverse interactions with Myo2-E1

```
TPR motif 1
                      Helix A          Helix B
                   -------------    -------------
                      *   *   *       *   *  *  * *
CeUNC-45      8    AEEIRDEGNAAVKDQDYIKADELYTEALQLTTDE
hGCUNC-45     21   VEQLRKEGNELFKCGDYGGALAAYTQALGLDATP
Hop           225  ALKEKELGNDAYKKKDFDTALKHYDKAKELDPTN
hPP5          28   AEELKTQANDYFKAKDYENAIKFYSQAIELNPSN
FKBP51        268  AAIVKEKGTVYFKGGKYMQAVIQYGKIVSWLEME

TPR motif 2
                      Helix 2A          Helix 2B
                   -------------     -------------
                      *   *   *       *      *  *
CeUNC-45      47   PVLYRNRAMARLKRDDFEGAQSDCTKALEFDGAD
hGCUNC-45     58   AVLHRNRAACHLKLEDYDKAETEASKAIEKDGGD
Hop           259  MTYITNQAAVYFEKGDYNKCRELCEKAIEVGREN
hPP5          62   AIYYGNRSLAYLRTECYGYALGDATRAIELDKKY
FKBP51        317  LAAFLNLAMCYLKLREYTKAVECCDKALGLDSAN

TPR motif 3
                      Helix 3A          Helix 3B
                   -------------    -------------
                      *   *           *   *  *  * *
CeUNC-45      81   VKALFRRSLAREQLGNVGPAFQDAKEALRLSPND
hGCUNC-45     92   VKALYRRSQALEKLGRLDQAVLDLQRCVSLEPKN
Hop           300  AKAYARIGNSYFKEEKYKDAIHFYNKSLAEHRTP
hPP5          96   IKGYYRRAASNMALGKFRAALRDYETVVKVKPHD
FKBP51        351  EKGLYRRGEAQLLMNEFESAKGDFEKVLEVNPQN
```

Figure 4A. UNC-45 homologs contain canonical TPR domains. The TPR domains of *C. elegans* UNC-45 (CeUNC-45; NP_497205) and human general cell UNC-45 (hGCUNC-45; NP_061141) were aligned with the Hsp90-binding TPR domains of human Hop (AAV38813), human protein phosphatase 5 (hPP5; NP_006238) and human FK506 binding protein 51 (FKBP51; NP_004108). The TPR consensus residues are indicated by asterisks.[33,52] Residues in grey backgrounds are involved in electrostatic and hydrogen bond interactions with Hsp90.[53] The broken lines represent the extent of helices A and B of each TPR motif.

mutant myosin.[54] Rng3p colocalizes with *myo2*-E1 mutant myosin at the cell division site in an F-actin-dependent manner. More importantly, Rng3p has been shown to be necessary for the formation of progenitor 'spots' that form the actomyosin ring assembly in interphase *S. pombe* cells.[57] Maintenance of the myosin-containing spots however, is independent of F-actin. While the actomyosin ring has a rapid turnover, the interphase spot does not, showing that this progenitor structure in the interphase is necessary to ensure proper assembly of the actomyosin ring and successful cell division. Recombinant full-length Rng3p or its UCS domain alone are necessary and sufficient to activate the actin-based motility of myosin in vitro and double its actin-activated Mg^{2+}-ATPase activity.[36] Whether Rng3p and other fungal UCS proteins require Hsp90 for their myosin-dependent functions is uncertain. However, in vivo, Swo1p (Hsp90 homolog in *S. pombe*) and Rng3p have been shown to be both

```
                                                      *
CeUNC-45     524 --AVISLAKTCKKFLLETEKYSVDIRRYACEGLSYLSLDADVKEWIVDDSLLLKALVLLA
Rng3p        322 LAFQYQLSQVIPTLKLLLQS--KVYDSVLLEALRQSSTLGPVKQLIADDSCLLNNLSKLL
CRO1         268 TTSIEDLSKRFTRMLLDEDE--IEHVQPSIEGLAYASLQPKVKESLSKDSKTLKRLVKAL
She4p        341 --SRRIMPKIENVNESAVKLEEVPKVEMSVEALAYLSLKASVKIMIRSNESFTEILLTMI
hSMUNC-45    506 --STEKLAKQCRKWLCNMSI-DTRTRRWAVEGLAYLTLDADVKDDFVQDVPALQAMFELA
hGCUNC-45    522 --GTLKLAKQCRKWLCNDQI-DAGTRRWAVEGLAYLTFDADVKEEFVEDAAALKALFQLS

CeUNC-45     582 KK--AGALCVYTLATIYANLSNAFEKPKVDEE-MVKLAQFAKHH--VPETHPKDTEEYVE
Rng3p        380 LDTNISPLDASSIATIIYNMCKFKITKSEHERELNQLRNMAEAS--KTIDYKEDETAPTE
CRO1         326 DEAPPRSPMIYGALSIFTNLTRYRPIETDEEKRIRQLKAYANAAGKLQQVDPLNEDEHVT
She4p        399 KS-QKMTHCLYGLLVIMANLSTLPEESNGSSQSINDLKNYADLK--GPGADKVGAEKESK
hSMUNC-45    563 KAGTSDKTILYSVATTLVNCTNSYDVKEVIPE-LVQLAKFSKQH--VPEEHPKDKKDFID
hGCUNC-45    579 RL--EERSVLFAVASALVNCTNSYDYEEPDPK-MVELAKYAKQH--VPEQHPKDKPSFVR

                                                                   ▼
CeUNC-45     637 KRVRALVEEGAV-PACVAVSKTES----KNALELIARSLLAFAEYEDLRGRIIAEGGTVL
Rng3p        438 RRIQKILEYDIL-SKLFSAAKHYN-----SLNGLLAMILVHMANYKLARRKLVQIGALKF
CRO1         386 ERCKRVFEAGLT-PVLIKQSKSGS----AASLALIISIIHALSTPPPLRGQLAQQGAVRL
She4p        456 EDILLFNEKYILRTELISFLKREMHNLSPNCKQQVVRVIYNITRSKNFIPQCISQGGTTI
hSMUNC-45    620 MRVKRLLKAGVI-SALACMVKADSAILTDQTKELLARVFLALCDNPKDRGTIVAQGGGKA
hGCUNC-45    634 ARVKKLLAAGVV-SAMVCMVKTESPVLTSSCRELLSRVFLALVEEVEDRGTVVAQGGGRA

CeUNC-45     692 CLR--LTKEASGEG-KIKAGHAIAKLGAKADPMISFPGQR---AYEVVKPLCDLLHPDVE
Rng3p        492 LTR--QCFIQTQDS---NAAFALAKILISVAPHSIFTKAFP--SNRAIHPMSKLLSTNSA
CRO1         441 LIAAWTALPETENGPKRAAAQALARILISTNPALVFGGTRPIPQSAAIRPLASILTPDPT
She4p        516 ILEYLANKQDIGEPIRILGCRALTRMLIFTNPGLIFKKYS---ALNAIPFLFELLPRSTP
hSMUNC-45    679 LIP--LALEGTDVG-KVKAAHALAKIAAVSNPDIAFPGER---VYEVVRPLVRLLDTQRD
hGCUNC-45    693 LIP--LALEGTDVG-QTKAAQALAKLTITSNPEMTFPGER---IYEVVRPLVSLLHLNCS

                                                                      *
CeUNC-45     746 G-------------KANYDSLLTLTNLASVS----DSIRGRILKEKAIPKIEEFWFMTDH
Rng3p        545 DT-------EYPILLGKFEVLLALTNLASHD----EESRQAIVQECWRELD-EL-IIETN
CRO1         501 AD-------RR-DLLPTFESLMALTNLASTD----DDTRKSIIR-TAWDDVEE-QLFNPN
She4p        573 VDDNPLHNDEQIKLTDNYEALLALTNLASSETSDGEEVCKHIVSTKVYWSTIENLMLGEN
hSMUNC-45    733 G-------------LQNYEALLGLTNLSGRS----DKLRQKIFKERALPDI-ENYMFENH
hGCUNC-45    747 G-------------LQNFEALMALTNLAGIS----ERLRQKILKEKAVPMI-EGYMFEEH

                                                       *
CeUNC-45     789 EHLRAAAAELLLNLLFFEKFYEETVAPGTDRLK----LWVLYSAEVEEERLSRASAAGFA
Rng3p        592 PLIQRATTELINNLSLSPYCLIKFIGDKDSDFENTR-LHIVLALSDTEDTPTRLAACGIL
CRO1         547 SRVCTAAVELVCNLVQDPEQTLALFGDGSPKAKNR-VKVIVALADAEDPKTRSAAGGALA
She4p        633 VPLQRSTLELISNMMSHPLTIAAKFFNLENPQSLRNFNILVKLLQLSDVESQRAVAAIFA
hSMUNC-45    775 DQLRQAATECMCNMVLHKEVQERFLADGNDRLK----LVVLLCGE-DDDKVQNAAAGALA
hGCUNC-45    789 EMIRRAATECMCNLAMSKEVQDLFEAQGNDRLK----LLVLYSGE-DDELLQRAAAGGLA

                                                        ▼
CeUNC-45     845 ILTED-ENACARIMDEIKSWPEVFKDIAMH-EDAETQRRGLMGIANIMHSSNKLC----S
Rng3p        652 VQITSVDEGCKKILSLQND-FNYIVRMLTD-QDEGIQHRGLVCICNIVYSKDQEIFN--K
CRO1         606 SLTGFDEVVRAVMGLERG--VEVVLGLCRD-EREDLRHRGAVVVRNMVFSEGEVGRLARG
She4p        693 NIATTIPLIAKELLTKKELIENAIQVFADQIDDIELRQRLLMLFFGLFEVIPDNG-----
hSMUNC-45    830 MLTAAHKKLCLKMTQVTTQWLEILQRLCLH-DQLSVQHRGLVIAYNLLAADAELA----K
hGCUNC-45    844 MLTSMRPTLCSRIPQVTTHWLEILQALLLS-SNQELQHRGAVVVLNMVEASREIA----S

CeUNC-45     899 EIVSSEVFRVLVAVTKLGTINQERAGSTEQAKRGLEAAEKFGLIKATDR-EIYERENQMS
Rng3p        707 FIKTPKAVETLRTYITK-------QAALKELQHEALVMIDSRLQGSK-------------
CRO1         663 KLVEGGAVEALMECAKG---------SKRREVVEVVVQAAEGLMGEGGK-----------
She4p        748 ---TNEVYPLLQENQKLKDALN---MSLKRDDSGPEFSAAIPVILAKIKV----------
hSMUNC-45    885 KLVESELLEILTVVGKQ-EPDEKKAEVVQTARECLIKCMDYGFIKPVS------------
hGCUNC-45    899 TLMESEMMEILSVLAKG-DHSP----VTRAAAACLDKAVEYGLIQPNQDGE---------

CeUNC-45     958 TIQE----
Rng3p            --------
CRO1             --------
She4p            --------
hSMUNC-45        --------
hGCUNC-45        --------
```

Figure 4B. Conserved mutations in the C-termini define the UCS domain of UNC-45 protein homologs. The UCS domains of animal UNC-45 proteins, *C. elegans* UNC-45 (CeUNC-45; NP_497205), human general cell UNC-45 (hGCUNC-45; NP_061141), human striated muscle UNC-45 (hSMUNC-45; AAI01064) and fungal UCS proteins, *S. pombe* ring assembly protein 3 (Rng3p; O74994), *P. anserina* CRO1 protein (CRO1; CAA76144) and *S. cerevisiae* She4p (She4p; NP_014678) were aligned. Mutations in homologous regions in CeUNC-45 and Rng3p that are used to define the UCS domain are indicated by asterisks and arrows respectively. A highly conserved region common to all UCS proteins is boxed.

required for Myo2 assembly in the contractile ring.[12] These observations suggest that some functional relationship exists between the *S. pombe* UCS protein and Hsp90.

The CRO1 protein of the filamentous fungus, *P. anserina*, is a 702-residue protein that is required for sexual sporulation.[8] GFP-tagging of the CRO1 protein reveals that it is a cytosolic protein expressed mainly at the beginning of the dikaryotic stage and at the time of ascospore maturation. The primary defect of null mutant allele of the gene, *cro1-1* is the inability to form septa between the daughter nuclei after mitotic division. The mutant also results in abortive meioses of resultant polyploid nuclei and lack of progression from the syncytial (vegetative) state to the cellular (sexual) state.[8] Unlike the wild type fungal filaments, disorganization of the actin prevents microtubule disassembly.

Conclusions and Future Work

The discovery of *C. elegans* UNC-45 and other UCS-domain proteins has led to the new research area of myosin-targeted chaperones. The complexity of myosin motor domain in terms of its structure and the multiplicity of its conformational states[59] suggest that it is a target for molecular chaperones. Evidence presented above confirms that the UCS proteins function as chaperones for myosins. In addition to assisting myosins to fold completely, UNC-45 and probably fungal UCS proteins may participate in myosin assembly, and modulate its contractile function and turnover.[35] It is notable that UNC-45 may target to myosin motors the participation of Hsp90, a chaperone whose interactions are necessary for the folding and functions of many different kinds of proteins. Much work still has to be done to fully understand the roles of UNC-45 and other UCS proteins.

Acknowledgements

This work was supported in part by grants from the National Institute of Health, the Muscular Aystropy Association, and from the Cecil H. and Ida M. Green Endowment.

References

1. Hutagalung AH, Landsverk ML, Price MG et al. The UCS family of myosin chaperones. J Cell Sci 2002; 115:3983-3990.
2. Yu QI, Bernstein SI. UCS proteins: Managing the myosin. Curr Biol 2003; 13:525-527.
3. Das AK, Cohen PTW, Barford D. The structure of the tetratricopeptide repeats of protein phosphatase 5, implications for TPR-mediated protein-protein interactions. EMBO J 1998; 17:1192-1199.
4. Price MG, Landsverk ML, Barral JM et al. Two mammalian UNC-45 isoforms are related to distinct cytoskeletal and muscle-specific functions. J Cell Sci 2002; 115:4013-4023.
5. Balasubramanian MK, McCollum D, Chang L et al. Isolation and characterization of new fission yeast cytokinesis mutants. Genetics 1998; 149:1265-1275.
6. Jansen RP, Dowzer C, Michaelis C et al. Mother cell-specific HO expression in budding yeast depends on the unconventional myosin myo4p and other cytoplasmic proteins. Cell 1996; 84:651-654.
7. Wendland B, McCaffery JM, Xiao Q et al. A novel fluorescence-activated cell sorter-based screen for yeast endocytosis mutants identifies a yeast homologue of mammalian eps15. J Cell Biol 1996; 135:1485-1500.
8. Berteaux-Lecellier V, Zickler D, Debuchy R et al. A homologue of the yeast SHE4 gene is essential for the transition between the syncytial and cellular stages during sexual reproduction of the fungus Podospora anserina. EMBO J 1998; 17:1248-1258.
9. Epstein HF, Thomson JN. Temperature-sensitive mutation affecting myofilament assembly in Caenorhabditis elegans. Nature 1974; 250:579-580.
10. Kachur T, Ao W, Berger J et al. Maternal UNC-45 is involved in cytokinesis and colocalizes with nonmuscle myosin in the early Caenorhabditis elegans embryo. J Cell Sci 2004; 117:5313-5321.
11. Barral JM, Hutagalumg AH, Brinker A et al. Role of myosin assembly protein UNC-45 as a molecular chaperone for myosin. Science 2002; 295:669-671.
12. Mishra M, D'souza VM, Chang KC et al. Hsp90 protein in fission yeast Swo1p and UCS protein Rng3p facilitate myosin II assembly and function. Eukaryot Cell 2005; 4:567-576.
13. Wesche S, Arnold M, Jansen RP. The UCS domain protein She4p binds to myosin motor domains and is essential for class I and class V myosin function. Curr Biol 2003; 13:715-724.

14. Yu Q, Hipolito LC, Kronert WA et al. Characterization and functional analysis of the Drosophila melanogaster unc-45 (dunc-45) gene. Mol Biol Cell 2003; 14:45.
15. Landsverk ML, Epstein HF. Genetic analysis of myosin II assembly and organization in model organisms. Cell Mol Life Sci 2005; 62:2270-2282.
16. Kuczmarski ER, Spudich JA. Regulation of myosin self-assembly: Phosphorylation of Dictyostelium heavy chain inhibits formation of thick filaments. Proc Natl Acad Sci USA 1980; 77:7292-7296.
17. Srikakulam R, Winkelmann DA. Myosin II folding is mediated by a molecular chaperonin. J Biol Chem 1999; 274:27265-27273.
18. Rayment I, Rypniewski WR, Schmidt-Base K et al. Three-dimensional structure of myosin subfragment-1: A molecular motor. Science 1993; 261:50-58.
19. Rayment I, Holden HM, Whittaker M et al. Structure of the actin-myosin complex and its implications for muscle contraction. Science 1993; 261:58-65.
20. Atkinson SJ, Stewart M. Expression in Escherichia coli of fragments of the coiled-coil rod domain of rabbit myosin: Influence of different regions of the molecule on aggregation and paracrystal formation. J Cell Sci 1991; 99:823-836.
21. Saraswat LD, Lowey S. Engineered cysteine mutants of myosin light chain 2: Fluorescent analogues for structural studies. J Biol Chem 1991; 266:19777-19785.
22. McNally EM, Goodwin EB, Spudich JA et al. Coexpression and assembly of myosin heavy chain and myosin light chain in Escherichia coli. Proc Natl Acad Sci 1988; 85:7270-7273.
23. Mitchell EJ, Karn J, Brown DM et al. Regulatory and essential light-chain-binding sites in myosin heavy chain subfragment-1 mapped by site-directed mutagenesis. J Mol Biol 1989; 208:199-205.
24. Trybus KM. Regulation of expressed truncated smooth muscle myosins. Role of the essential light chain and tail length. J Biol Chem 1994; 269:20819-20822.
25. Sweeney HL, Rosenfeld SS, Brown F et al. Kinetic tuning of myosin via a flexible loop adjacent to the nucleotide binding pocket. J Biol Chem 1998; 273:6262-6270.
26. Wang F, Chen L, Arcucci O et al. Effect of ADP and ionic strength on the kinetic and motile properties of recombinant mouse myosin V. J Biol Chem 2000; 275:4329-4335.
27. Kinose F, Wang SX, Kidambi US. Glycine 699 is pivotal for the motor activity of skeletal muscle myosin. J Cell Biol 1996; 134:895-909.
28. Chow D, Srikakulam R, Chen Y et al. Folding of the striated muscle myosin motor domain. J Biol Chem 2002; 277:36799-36807.
29. Srikakulam R, Winkelmann DA. Chaperone-mediated folding and assembly of myosin in striated muscle. J Cell Sci 2004; 117:641-652.
30. Chen S, Smith DF. Hop as an adaptor in the heat shock protein 70 (hsp70) and hsp90 chaperone machinery. J Biol Chem 1998; 273:35194-35200.
31. Morishima Y, Murphy PJ, Li DP et al. Stepwise assembly of a glucocorticoid receptor.hsp90 heterocomplex resolves two sequential ATP-dependent events involving first hsp70 and then hsp90 in opening of the steroid binding pocket. J Biol Chem 2000; 275:18054-18060.
32. Young JC, Moarefi I, Hartl FU. Hsp90: A specialized but essential protein-folding tool. J Cell Biol 2001; 154:267-273.
33. Blatch GL, Lassle M. The tetratricopeptide repeat: A structural motif mediating protein-protein interactions. BioEssays 1999; 21:932-939.
34. Smith DF. Tetratricopeptide repeat co-chaperones in steroid receptor complexes. Cell Stress Chap 2004; 9:109-121.
35. Hoppe T, Cassata G, Barral JM et al. Regulation of the myosin-directed chaperone UNC-45 by a novel E3/E4-multiubiquitylation complex in C. elegans. Cell 2004; 118:337-349.
36. Lord M, Pollard TD. UCS protein Rng3p activates actin filament gliding by fission yeast myosin-II. J Cell Biol 2004; 167:315-321.
37. Millson SH, Truman AW, King V et al. A two-hybrid screen of the yeast proteome for Hsp90 interactors uncovers a novel Hsp90 chaperone requirement in the activity of a stress-activated mitogen-activated protein kinase, Slt2p (Mpk1p). Eukaryot Cell 2005; 4:849-860.
38. Bose S, Weikl T, Bugl H et al. Chaperone function of Hsp90-associated proteins. Science 1996; 274:1715-1717.
39. Freeman BC, Toft DO, Morimoto RI. Molecular chaperone machines: Chaperone activities of the cyclophilin Cyp-40 and the steroid aporeceptor-associated protein p23. Science 1996; 274:1718-1720.
40. Kimura Y, Rutherford SL, Miyata Y. Cdc37 is a molecular chaperone with specific functions in signal transductions. Genes Dev 1997; 11:1775-1785.
41. Venolia L, Waterston RH. The unc-45 gene of Caenorhabditis elegans is an essential muscle-affecting gene with maternal expression. Genetics 1990; 126:345-353.
42. Barral JM, Bauer CC, Ortiz I et al. Unc-45 mutations in Caenorhabditis elegans implicate a CRO1/She4p-like domain in myosin assembly. J Cell Biol 1998; 143:1215-1225.

43. Waterston RH. Muscle. In: Wood WB, ed. The Nematode Caenorhabditis elegans. Cold Spring Harbor, New York: Cold Spring Harbor Laboratory, 1988:281-335.
44. Waterston RH, Thomson JN, Brenner S. Mutants with altered muscle structure in Caenorhabditis elegans. Dev Biol 1980; 77:271-302.
45. Schachat F, Harris HE, Epstein HF. Two homogeneous myosins in body-wall muscle of Caenorhabditis elegans. Cell 1977; 10:721-728.
46. Miller DM, Ortiz L, Berliner GC et al. Differential localization of two myosins within nematode thick filaments. Cell 1983; 34:477-490.
47. Waterston RH. The minor myosin heavy chain, mhcA, of Caenorhabditis elegans is necessary for the initiation of thick filament assembly. EMBO J 1989; 8:3429-3436.
48. Ardizzi JP, Epstein HF. Immunochemical localization of myosin heavy chain isoforms and paramyosin in developmentally and structurally diverse muscle cell types of the nematode Caenorhabditis elegans. J Cell Biol 1987; 105:2763-2770.
49. Venolia L, Ao W, Kim S et al. unc-45 gene of Caenorhabditis elegans encodes a muscle-specific tetratricopeptide repeat-containing protein. Cell Mot Cytoskeleton 1999; 42:163-177.
50. Ao W, Pilgrim D. Caenorhabditis elegans UNC-45 is a component of muscle thick filaments and colocalizes with myosin heavy chain B, but not myosin heavy chain A. J Cell Biol 2000; 148:375-384.
51. Guo S, Kemphues KJ. A nonmuscle myosin required for embryonic polarity in Caenorhabditis elegans. Nature 1996; 382:455-458.
52. Sikorski RS, Boguski MS, Goebl M. A repeating amino acid motif in CDC23 defines a family of proteins and a new relationship among genes required for mitosis and RNA synthesis. Cell 1990; 60:307-317.
53. Scheufler C, Brinker A, Bourenkov G et al. Structure of TPR domain-peptide complexes: Critical elements in the assembly of the Hsp70-Hsp90 multichaperone machine. Cell 2000; 101:199-210.
54. Wong KCY, Naqvi NI, Lino Y et al. Fission yeast Rng3p: An UCS-domain protein that mediates myosin II assembly during cytokinesis. J Cell Sci 2000; 113:2421-2432.
55. Young JC, Barral JM, Hartl FU. More than folding: Localized functions of cytosolic chaperones. TIBS 2003; 28:541-547.
56. Toi H, Fujimura-Kamada K, Irie K et al. She4p/Dim1p interacts with the motor domain of unconventional myosins in the budding yeast, Saccharomyces cerevisiae. Mol Biol Cell 2003; 14:2237-2249.
57. Wong KCY, D'souza VM, Naqvi NI et al. Importance of a myosin II-containing progenitor for actomyosin ring assembly in fission yeast. Curr Biol 2002; 12:724-729.
58. Pratt WB, Toft DO. Steroid receptor interactions with heat shock protein and immunophilin chaperones. Endocr Rev 1997; 18:306-360.

CHAPTER 7

The Roles of GroES as a Co-Chaperone for GroEL

Han Liu and Peter A. Lund*

Abstract

GroES works with the essential chaperone GroEL to mediate the folding of certain proteins from an unfolded or partially folded state. These two proteins form the only essential chaperone machine in *E. coli*. Both proteins have seven-fold symmetry. GroES acts by binding to one end of the GroEL complex in the presence of nucleotide. In doing this, it has several roles. It displaces bound substrate protein from GroEL into the folding cavity within the GroEL complex, and caps it while the protein folds. It also helps mediate the allosteric transitions that the GroEL complex undergoes during the course of its reaction cycle. A key part of the GroES co-chaperone is an extended loop of amino-acids that is highly mobile when the protein is free but becomes ordered on binding to GroEL, and the interaction between this mobile loop and GroEL helps define both the strength of the binding and the speed with which the chaperone machine passes through its cycle.

A Note on Nomenclature

Many different names have been used to refer to the co-chaperone GroES, its chaperone partner GroEL, and their many homologues. Strictly, the names GroEL and GroES refer only to the two proteins found in *E. coli* which together form the GroEL/GroES chaperone complex. The GroEL protein is responsible for the binding and folding of substrate proteins, and is thus a true molecular chaperone. GroES has no independent role as a chaperone, but acts on GroEL to modify its properties, and is hence correctly referred to as a co-chaperone. GroEL and its many homologues are also called the chaperonins,[1] a term which should not be applied (but sometimes is) to chaperones in general. GroES is therefore often referred to as a co-chaperonin. Homologues of the GroEL and GroES proteins go by various names, depending on the organism or organelle in which they are found. A general term for GroEL and GroES homologues is Cpn60 and Cpn10 proteins (for chaperonin).[2] The homologues in mitochondria are usually referred to as the Hsp60 and Hsp10 proteins. The homologues in chloroplasts are usually referred to as Cpn60 and Cpn10 or Cpn20 proteins although the chloroplast Cpn60 protein was originally referred to as "Rubisco binding protein".[1] In the text that follows, GroEL and GroES will be used strictly to refer to the *E. coli* proteins.

*Corresponding Author: Peter A. Lund—School of Biosciences, University of Birmingham, Birmingham B15 2TT, U.K. Email: p.a.lund@bham.ac.uk

Networking of Chaperones by Co-Chaperones, edited by Gregory L. Blatch.

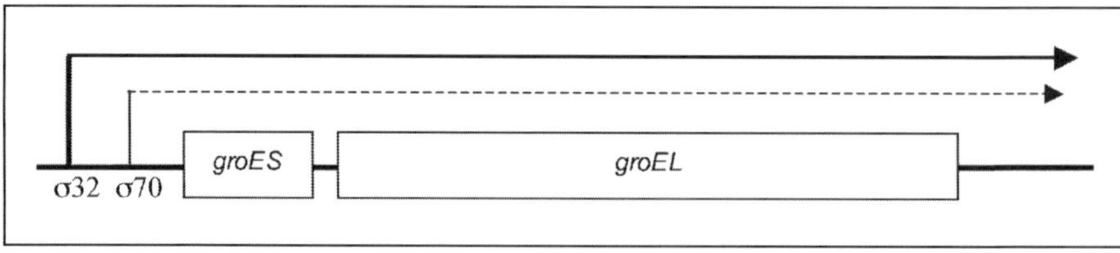

Figure 1. Organisation and transcription of the *groE* operon in *E. coli*.

The in Vivo Roles of the GroEL and GroES Proteins

The *groE* locus was first discovered in a screen for mutants of *E. coli* that failed to plate bacteriophage lambda. It was subsequently noted that mutants at this locus were temperature sensitive and failed to plate several other bacteriophages, including ones quite distinct from bacteriophage lambda, pointing to a protein with a rather broad function.[3] Subsequent genetic analysis showed there to be two genes present at the *groE* locus,[4] which were named *groEL* (for large) and *groES* (for small). The proteins they encode have molecular masses of 57.3 kDa and 10.3 kDa respectively, and it was soon established that both proteins assembled into high molecular weight complexes, with fourteen sub-units in the GroEL complex and seven in the GroES complex.[5,6] Subsequent experiments showed that both of these proteins were essential for *E. coli* growth.[7] The *groES* and *groEL* genes are organized in an operon in that order, transcribed predominantly from a sigma-32 dependent promoter, although with some expression also from a sigma-70 dependent promoter (Fig. 1).[8] Sigma-32 is the subunit of RNA polymerase that directs transcription of the heat shock genes of *E. coli*, and expression of the *groE* operon is highly heat shock inducible. Early genetic evidence showed that the GroEL and GroES complexes associated,[9] and this was confirmed biochemically, in studies which also showed the association required the presence of nucleotide.[6]

A significant advance was made in the understanding of these proteins when it was discovered that the GroEL protein of *E. coli* had a homologue in plant chloroplasts which was involved in assisting the large sub-unit of Rubisco to reach its fully folded form.[1] On this basis it was proposed that both these proteins had a role in helping other proteins to reach their active form, an idea that was somewhat heretical as it appeared to contradict the previously established fact that proteins are most thermodynamically stable in their folded state, and hence should (and under defined in vitro conditions, demonstrably do) fold spontaneously.[10] However, it was subsequently demonstrated that certain proteins folded more efficiently in the presence of the GroEL and GroES proteins. This was first shown in vivo using a bacterial Rubisco as a substrate. This protein folds very poorly when expressed in *E. coli*, but if the levels of both the GroEL and GroES proteins were increased by expression from a strong promoter, folding could be markedly improved.[11] Significantly, it was shown that both GroEL and GroES had to be expressed for this improved folding to occur.

Much work on the GroEL/GroES system has established that together these two proteins form a chaperone machine, which can assist the refolding of a variety of proteins from denatured or partially unfolded conformations to their folded, active states. The GroEL protein provides the environment where folding occurs (often referred to as the "Anfinsen cage"[12]), but the GroES protein also has a number of key roles in the efficient functioning of this chaperone machine, as will be discussed in detail below.

In *E. coli*, and by extension in other bacteria, the GroEL/GroES chaperone has a relatively small number of obligate substrates—i.e., proteins which rely on the GroEL/GroES chaperone machine for folding and which, if it is not present, fail to fold. In total, 89 such substrate proteins have been identified to date, of which 13 are known to be essential for *E. coli* to grow, thus explaining the essential nature of the *groEL* and *groES* genes.[13] Although it has not been formally proven that every obligate GroEL substrate in *E. coli* also requires GroES, this is true

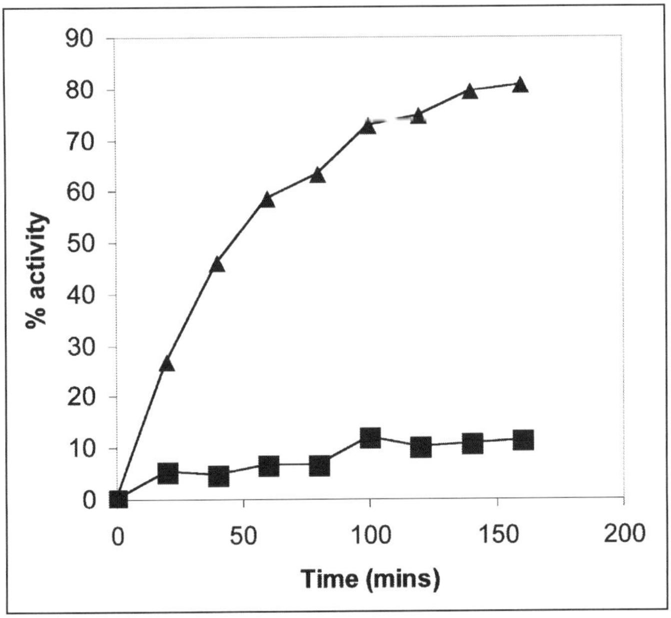

Figure 2. Refolding of mitochondrial malate dehydrogenase from the denatured state, in the presence (filled triangles) or absence (filled squares) of GroES. The refolding mix also contained GroEL and ATP.

for all cases tested in vitro. A major focus of research on these proteins in the last fifteen years has been trying to establish the precise mechanism by which they both function, and to relate this to the remarkable structures of the complexes that the two proteins form. This will be discussed in the sections that follow.

The Roles of GroES in the Chaperonin Mechanism

Early Experiments on the GroEL/GroES Chaperone Machine

Initial in vitro evidence for the chaperone activity of GroEL and GroES came from experiments where bacterial Rubisco (ribulose bisphosphate carboxylase-oxygenase) was denatured, and then renatured in the presence of purified GroEL, GroES, and ATP. Under these conditions, the yield of active Rubisco was shown to increase to a maximum of 80%.[14,15] This was completely dependent on the presence of both GroEL and GroES; in the absence of either of these two proteins, or of ATP, no active protein was formed. Many subsequent studies of this type using a variety of protein substrates investigated the function of GroEL/GroES in vitro. The results of these showed that GroEL/GroES acts by suppressing aggregation and facilitating folding. Examples of such studies are described in references 16-20, and an example from the authors' laboratory is shown in Figure 2. However, the need for the co-chaperone GroES in this reaction was variable: in some cases, protein folding was shown to require both GroEL and GroES, whereas in others it could be shown that the presence of GroEL alone was sufficient for folding and the co-chaperone was not required. A detailed investigation into this phenomenon showed that the requirement for GroES frequently depended on the precise nature of the conditions employed for protein refolding.[21] Under stringent conditions, when folding of the protein in the absence of chaperones was negligible, GroES was shown to be required for the chaperone complex to act, whereas if the protein was able to fold up to some extent in the absence of the complete chaperone system, GroES became less important in the reaction. However, even in cases where GroES was not absolutely required for folding, it was shown to speed up the refolding reaction.

It was unclear from these early studies exactly what the role of the GroES co-chaperonin in the reaction mechanism of the GroEL/GroES machine might be. The current view of the mechanism of GroEL/GroES is that it acts by encapsidating unfolded or partially proteins in the central cavity of one of the GroEL rings, capped by bound GroES protein. This encapsidation gives the protein time to fold without interacting with other unfolded proteins, and hence reduces the risk of protein aggregation. Detailed insights into this process had to await two major developments: a description of the different structures of the GroEL/GroES machine as it progresses through the protein refolding cycle, and a more detailed understanding of the individual steps of the chaperone-mediated protein refolding cycle itself. These will be discussed in the next two sections.

The Structure of GroES and the GroEL/GroES Complex

The structures of GroEL, GroES, and the GroEL/GroES-nucleotide complex have been extensively studied with electron microscopy[22-28] and X-ray crystallography.[29-32] More recently, NMR spectroscopy has been used to investigate the structure of GroES either free in solution or bound to GroEL.[33] GroEL consists of 14 identical subunits that form two rings stacked back to back. Each subunit folds into 3 distinctive domains: an apical domain that binds substrates and GroES,[34] an equatorial domain that contributes to inter-ring and most intra-ring contacts and also contains the nucleotide binding site, and a flexible intermediate domain that connects the other two domains. GroES consists of 7 identical subunits that form a dome-shaped structure (Fig. 3). Each subunit is composed of an irregular anti-parallel β-barrel and two β-hairpin loops (one upward and the other downward). The upward loops from the 7 subunits collectively form the roof of the dome. The downward loop contains about 20 amino-acid residues and was largely undefined in the original structure of GroES, showing that it has high mobility. This mobile loop modulates the interaction between GroES and GroEL, and becomes structurally ordered upon binding of GroES to GroEL, when a highly conserved hydrophobic tripeptide (I25, V26, and L27 in GroES) makes the major physical contacts with GroEL, principally with residues found in two short α-helices at the top of the GroEL apical domain;[31,32,35,36] see Figure 4. Mutations in the mobile loop region of GroES result in altered flexibility of the loop and changed affinity to its GroEL partner, which can affect the function of GroEL.[36] The GroEL/GroES-$(ADP)_7$ complex has a "bullet" shape, in which GroES binds to one end of the double ring of GroEL, thus covering the cavity in one of the two GroEL rings (Figs. 5, 6).[26,32] The ring that is bound to GroES is termed the cis ring, and the opposite ring that is free of GroES is termed the trans ring. The binding of GroES to GroEL is accompanied by major changes in the conformation of the sub-units in the cis ring of GroEL, and minor ones in the trans ring. In the cis ring, these involve large rigid body motions of the apical domain, which lead to an approximate doubling of the volume of the cavity inside the GroEL ring,[26,32] relative to the size of the cavity in the absence of bound GroES. This increases the maximum theoretical size of unfolded proteins that can be encapsidated in the cavity.

Under certain conditions, GroES can bind to both ends of the GroEL complex, forming a structure whose shape is reminiscent of an American football. Controversy has long existed over whether the "bullet" or "football" structures are the more important in the folding cycle of proteins bound to GroEL.[38-40] The consensus is now strongly in favour of the bullets, but it is fair to say that the majority of in vitro experiments done on the GroEL/GroES system are under conditions which are far from physiological, so it would be premature to completely rule out a role for a "double-ended" GroEL/GroES complex.

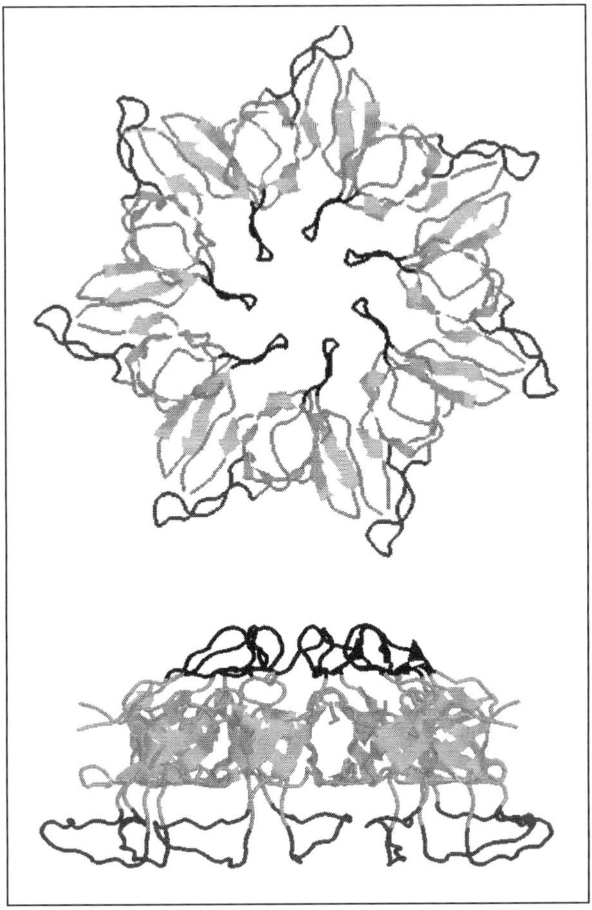

Figure 3. Ribbon representation of top view (above) and side view (below) of the GroES heptameric ring, as it occurs in a complex with GroEL and nucleotide (drawn using RasMol).[76] The "flexible loop" domains can be seen at the bottom of the ring, and the "lid" domain at the top. Both domains are coloured darker than the rest of the protein.

The Roles of GroES in the Chaperonin Reaction Cycle

The complexity of the structures of GroEL and GroES reflect the sophistication of this chaperone system. GroEL and GroES chaperone protein folding by providing a protected environment (the GroEL cavity, capped by the GroES co-chaperone) for proteins to fold, where they are shielded from other folding proteins (with which they might otherwise aggregate) long enough for the hydrophobic side-chains on the folding protein to become buried.[41] In addition, the inside of the cavity is lined with hydrophilic side chains, and it has been shown that encapsidation accelerates the rate of folding, at least for Rubisco, possibly due to these side chains favouring the burial of hydrophobic side chains in the folding protein.[42] This mechanism is sometimes referred to as the cis folding mechanism, as protein folding takes place on the ring that also binds to GroES. The reaction cycle of the cis folding mechanism is explained briefly here; for more detailed descriptions, these excellent reviews are recommended by the authors.[43-46] The reaction cycle starts with the binding of unfolded,

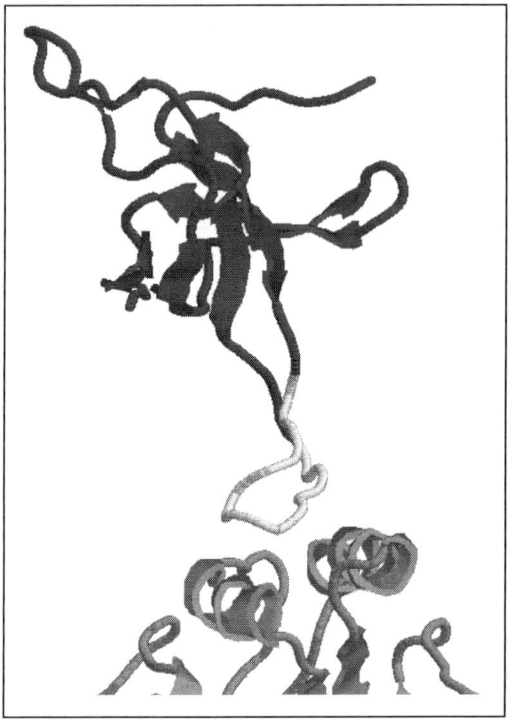

Figure 4. Ribbon representation of GroES, showing the mobile loop (light shading) interacting with the top of the apical domain of GroEL (drawn using RasMol).

partially folded, or misfolded substrate polypeptide, through hydrophobic interactions with residues in the apical domain of GroEL, to one ring of a GroEL complex, which thus becomes the cis ring. The subsequent binding of ATP to this complex initiates the movements of the apical domains of GroEL which are completed, and stabilized, by the binding of GroES.[26,32,47-49] The binding of ATP is positively cooperative within a single ring but negatively cooperative between rings; under cellular conditions only the cis ring binds ATP.[50] The twisting and elevation of the apical domains caused by GroES binding leads to burial of the hydrophobic patches on GroEL to which the substrate initially was bound, and GroES binding hence displaces the substrate into the GroEL cavity, where folding in a protected environment can now take place. The time available for this is set by the time taken for the cooperative hydrolysis of ATP by the GroEL subunits in the cis ring, which is several seconds. During this time, conformational changes in the trans ring caused by the binding of substrate, ATP, and GroES to the cis ring mean that the trans ring is not competent to bind substrate. The hydrolysis of ATP to ADP in the cis ring weakens the binding of GroES, and furthermore once ATP has been hydrolysed to ADP, the trans ring now becomes competent to bind another unfolded substrate protein and ATP. This causes the release of GroES from the cis ring, and the protein is free to diffuse away from the cavity.[49,50] This process is followed by binding of GroES to the trans ring.[51] Thus, the trans ring now has substrate, ATP, and GroES bound, and so has become the new cis ring for the next stage of the folding cycle. For this reason the GroEL/GroES system has sometimes been referred to as a "two stroke motor" as binding and folding proceeds at alternating ends of the complex.[52] However, if the substrate has failed to fold, it is likely to be recaptured by the hydrophobic regions on the

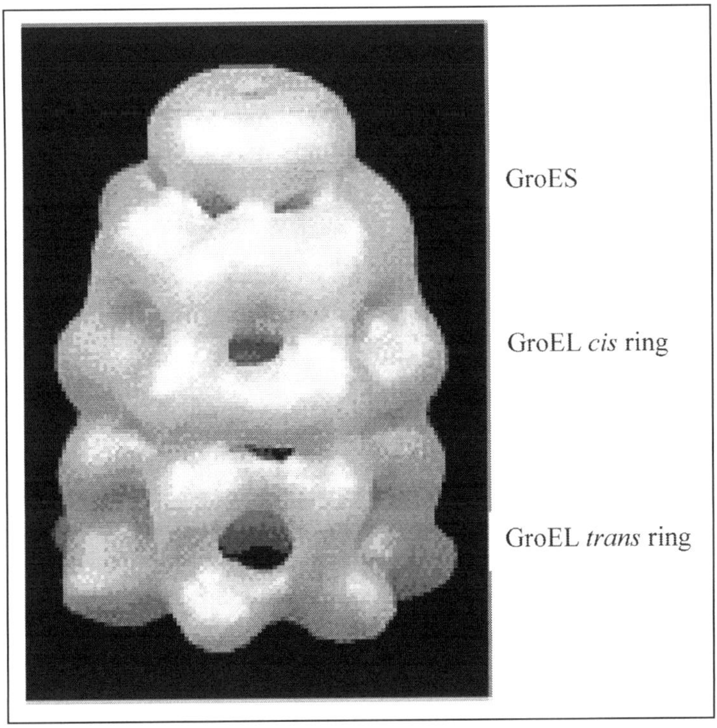

Figure 5. 30Å resolution structure of the GroEL/GroES-ADP$_7$ complex, generated from cryo-electron microscopy data, showing the capping of one end of the GroEL protein by GroES.[26] The significant difference in structure between the cis and trans rings is clearly visible. Reprinted from Roseman AM et al, Cell 87:241-51.[26] ©1996, with permission from Elsevier.

apical domain of GroEL, and go through another round of the reaction cycle. The entire cycle is shown diagrammatically in Figure 7.

In this cis folding model, the co-chaperone GroES assists the function of GroEL in several ways. (1) The binding of GroES provides a lid to the GroEL cavity, preventing substrate proteins from diffusing out of from the GroEL cavity. (2) The movement which it causes in the apical domains of GroEL results in the volume of the cavity in the cis ring of GroEL increasing approximately two-fold from 85,000 Å^3 to 175,000 Å^3.[32] The theoretical maximum size of a globular protein that could be encapsidated in this cavity is around 70kDa, even if fully folded; the majority of in vivo GroEL substrates are in fact found to be <60kDa.[53] (3) GroES, which has a hydrophilic inner surface, may help to establish a hydrophilic environment in the cis cavity that will favor the folding/refolding process of substrate proteins. Interestingly, in an experiment where the *groES* and *groEL* genes were allowed to evolve to improve their ability to fold a particular substrate (GFP), one of the mutations found was on the inner surface of the GroES dome, and led to an increase in hydrophilicity.[54] (4) GroES controls the pace of the reaction through its influence on the binding and hydrolysis of ATP by GroEL subunits, an effect exerted both on the cis ring, where it increases the cooperativity of ATP binding and hydrolysis and on the trans ring,[55] where it promotes the allosteric transformation of the trans ring to a form with a low affinity for nucleotide and unfolded protein.[50,56]

Polypeptides larger than 60 kDa that cannot be encapsidated beneath GroES have nevertheless been shown to associate with GroEL both in vivo and in vitro.[53,57-60] Moreover, the

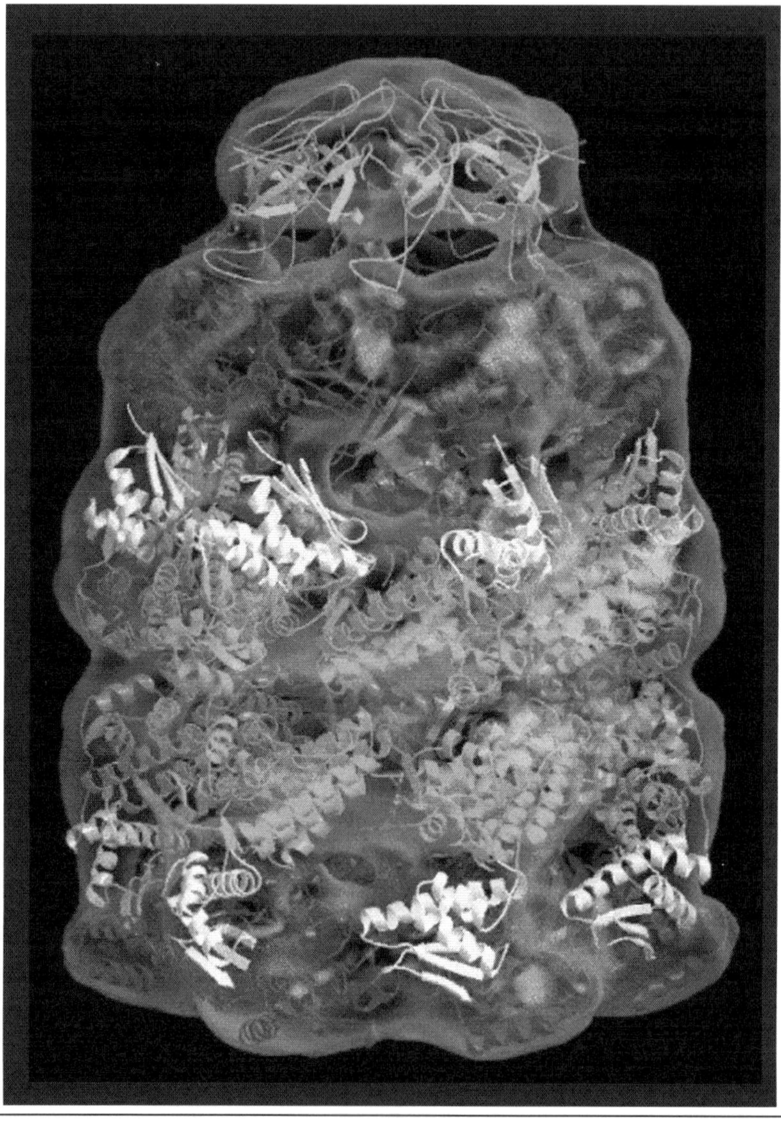

Figure 6. 12.5Å resolution structure generated from cryo-electron microscopy data of a GroEL-ATP$_7$-GroES-ADP$_7$ complex, with the atomic resolution structure docked in reference 28. The cryo-EM image is translucent, and the atomic structure is shown in ribbon conformation. Reprinted from Ranson NA et al; Cell 107:869-879.[28] ©2001, with permission from Elsevier.

refolding of denatured aconitase (molecular weight of 82 kDa) has been shown to require the assistance of GroEL and GroES in vitro.[61] This has been shown to be via a trans folding mechanism, in which the substrate is bound to the open trans ring of GroEL, and later released into the bulk solution to fold. This trans folding mechanism was investigated both in vivo and in vitro using trans-only constructs of GroEL/GroES, where GroES could not be fully released

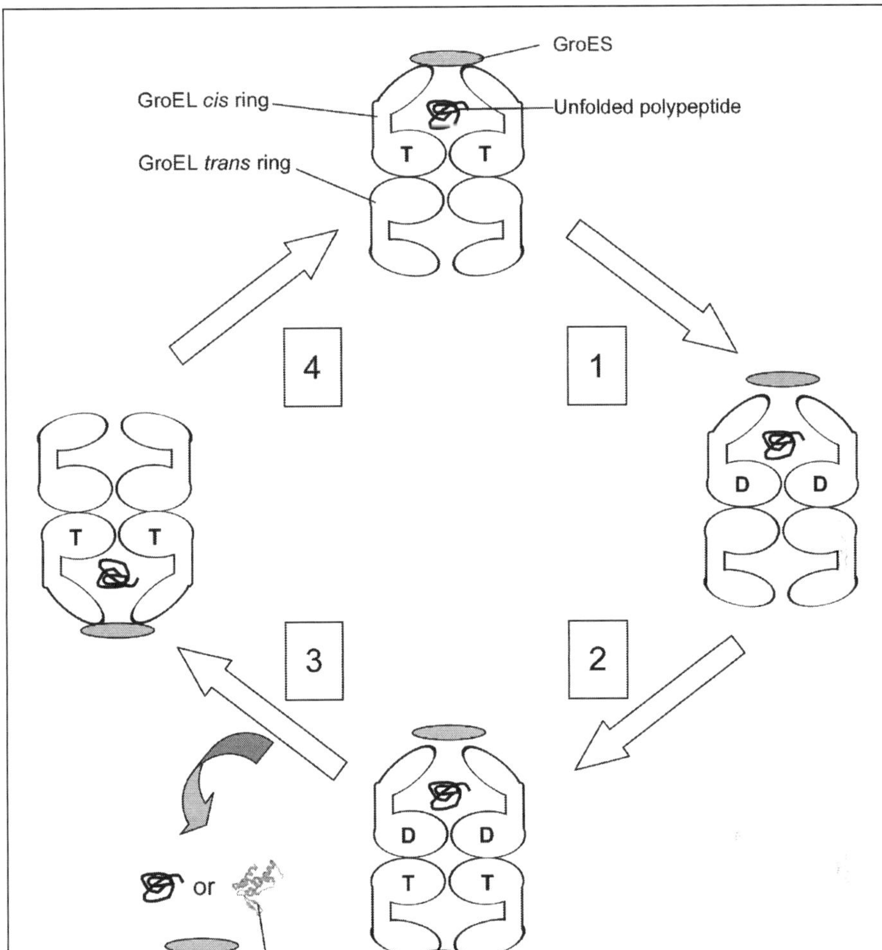

Figure 7. The GroEL/GroES reaction cycle. Stage 1 shows the hydrolysis of ATP bound in the cis ring of GroEL to ADP, which weakens the GroES-GroEL interaction. Stage 2 shows the binding of ATP and substrate to the trans ring, which leads in stage 3 to the discharge of GroES, ADP, and substrate from the cis ring. In stage 3, GroES also binds to the trans ring which hence becomes the new cis ring for the next round of the cycle. D: ADP; T: ATP.

from the GroEL/GroES complex and where no substrate could bind to the cis ring.[62] Remarkably, the results revealed that the trans-only construct could assist the refolding of not only denatured aconitase, but also of some proteins that were normally encapsulated in the cis ring, and it could also support the growth of cells lacking wild type GroEL/GroES, although growth was poor.[61] Thus encapsidation of bound protein in the GroEL cavity by GroES is not an absolute requirement for the GroEL/GroES system to act, but GroES is essential for both cis and trans folding mechanisms.

Mechanistic Insights from the Properties of Other Co-Chaperonins

The *groES* and *groEL* genes were originally discovered through selection for mutants that showed loss of ability to grow bacteriophage lambda. Subsequent studies with several other bacteriophage have shown that both GroES and GroEL are required for phage growth, usually because of a requirement for the assembly of either the head or tail structures. An interesting exception, however, is the T4 bacteriophage, which requires GroEL but which encodes its own co-chaperonin, known as gp31. The gp31 protein is required specifically for folding the T4 tail protein gp23.[63-65] The gp23 protein cannot be refolded by the GroEL/GroES complex, so T4 phage which are defective in gp31 production fail to form plaques on *E. coli*. Remarkably, gp31 can substitute for GroES not only in the folding of the gp23 proteins but also in the folding of other proteins and in normal bacterial growth; thus, for example, strains where *groES* is deleted can grow successfully as long as gp31 is expressed.[66,67] Thus gp31 is a co-chaperone for GroEL that is able to affect the properties of its partner chaperone in ways which are different to those of the normal co-chaperone, GroES. It has a structure which is similar to that of GroES, with a flexible loop domain which interacts with GroEL and a dome shape, even though the amino-acid homology between the two proteins is only 14%.[68] Recent results show two particular differences in the interaction of gp23 with the GroEL/gp31 complex compared with GroEL/GroES. First, gp23 can be bound by the trans ring in a GroEL/gp23 complex, but not by the trans ring in a GroEL/GroES complex. This shows that the nature of the co-chaperonin that is bound to the cis ring of GroEL exerts its allosteric effect all the way to the opposite end of the GroEL complex. The structural basis for the transmission of allosteric information, and of the difference that results in the trans ring, is not yet understood. Second, gp23 can be enclosed in the cavity on the cis ring of the GroEL/gp31 complex, but not the GroEL/GroES complex. This confirms that the size of the cavity, which is defined by the co-chaperonin, is critical in determining the upper size limit of substrates for the GroEL/GroES complex.[69]

Mitochondria contain a chaperone-co-chaperone pair known as Hsp60 and Hsp10. Unlike GroEL, Hsp60 can function as a single ring, although whether it is always a single ring inside mitochondria is not clear.[70-72] How then does it complete the chaperonin reaction cycle, given that there is no trans-ring to provide an allosteric signal to eject bound GroES and substrate protein? The answer lies in the nature of the interaction between Hsp10 and Hsp60. The complex formed between Hsp10 and Hsp60 in the presence of ADP is very weak,[71] so that Hsp10 can dissociate from Hsp60 as soon as the bound ATP has been hydrolysed. A critical component of the interaction is defined by the properties of the mobile loop. It has been shown that Hsp10 has a higher affinity for GroEL than does GroES; presumably this higher affinity is required in order to enable any interaction at all with Hsp60.[73] Furthermore, although GroEL can function in vivo and in vitro with Hsp10 substituting for GroES, GroES is unable to function with Hsp60. However, by swapping the mobile loop regions between the two proteins, a variant of GroES that can now act with Hsp60 can be created. Thus, the binding of GroES and other co-chaperonins to their chaperonin partners is precisely modulated by the nature of the mobile loop, and indeed it has been shown that the rate of the transition of the mobile loop between its disordered and ordered states is key in determining the speed with which chaperonins pass through the reaction cycle.[74] This result combines with the many others discussed above to show that the co-chaperonins, far from being inert "caps" to the chaperonin folding chamber, play a key role in many aspects of the complete chaperonin reaction cycle.

Future Research Directions

GroES is probably one of the best understood co-chaperones; in particular, in the relationship of its structure to its function. Kinetic analyses of the precise points in the GroEL/GroES reaction cycle where GroES has its effects are now underway in several laboratories, and a detailed structural description of exactly how it interacts with and modulates the nature of GroEL intermediates cannot be far behind. But it must be remembered that the vast majority of work in this field has focused on the model GroEL/GroES system of *E. coli*. As demonstrated in the previous section, our understanding of homologues of this system in other organisms and organelles is still relatively poor, and further effort is needed to establish a more general understanding of the chaperonins and their co-chaperonins. GroES homologues appear in some unexpected guises, too, and our understanding of these is very limited indeed. The most intriguing is perhaps the identification of "early pregnancy factor"—a protein that appears in the sera of animals, including humans, which contain a fertilized embryo even before its implantation—as an extracellular form of the mitochondrial GroES homologue, Hsp10.[75] There is currently no obvious explanation for this in terms of the co-chaperone properties of this molecule, and this fact alone suggests that a great deal more remains to be learned about the properties of this family of co-chaperones.

References

1. Hemmingsen SM, Woolford C, van der Vies SM et al. Homologous plant and bacterial proteins chaperone oligomeric protein assembly. Nature 1988; 333:330-334.
2. Coates AR, Shinnick TM, Ellis RJ. Chaperonin nomenclature. Mol Microbiol 1993; 8:787.
3. Friedman DI, Olson ER, Georgopoulos C et al. Interactions of bacteriophage and host macromolecules in the growth of bacteriophage lambda. Microbiol Rev 1984; 48:299-325.
4. Tilly K, Murialdo H, Georgopoulos C. Identification of a second Escherichia coli groE gene whose product is necessary for bacteriophage morphogenesis. Proc Natl Acad Sci USA 1981; 78:1629-1633.
5. Hohn T, Hohn B, Engel A et al. Isolation and characterization of the host protein groE involved in bacteriophage lambda assembly. J Mol Biol 1979; 129:359-373.
6. Chandrasekhar GN, Tilly K, Woolford C et al. Purification and properties of the groES morphogenetic protein of Escherichia coli. J Biol Chem 1986; 261:12414-12419.
7. Fayet O, Ziegelhoffer T, Georgopoulos C. The groES and groEL heat shock gene products of Escherichia coli are essential for bacterial growth at all temperatures. J Bacteriol 1989; 171:1379-1385.
8. Zhou YN, Kusukawa N, Erickson JW et al. Isolation and characterization of Escherichia coli mutants that lack the heat shock sigma factor sigma 32. J Bacteriol 1988; 170:3640-3649.
9. Tilly K, Georgopoulos C. Evidence that the two Escherichia coli groE morphogenetic gene products interact in vivo. J Bacteriol 1982; 149:1082-1088.
10. Anfinsen CB. Principles that govern the folding of protein chains. Science 1973; 181:223-230.
11. Goloubinoff P, Gatenby AA, Lorimer GH. GroE heat-shock proteins promote assembly of foreign prokaryotic ribulose bisphosphate carboxylase oligomers in Escherichia coli. Nature 1989; 337:44-47.
12. Ellis RJ. Molecular chaperones. Opening and closing the Anfinsen cage. Curr Biol 1994; 4:633-635.
13. Kerner MJ, Naylor DJ, Ishihama Y et al. Proteome-wide analysis of chaperonin-dependent protein folding in Escherichia coli. Cell 2005; 122:209-220.
14. Goloubinoff P, Christeller JT, Gatenby AA et al. Reconstitution of active dimeric ribulose bisphosphate carboxylase from an unfolded state depends on two chaperonin proteins and Mg-ATP. Nature 1989; 342:884-889.
15. Viitanen PV, Lubben TH, Reed J et al. Chaperonin-facilitated refolding of ribulosebisphosphate carboxylase and ATP hydrolysis by chaperonin 60 (groEL) are K+ dependent. Biochemistry 1990; 29:5665-5671.
16. Martin J, Langer T, Boteva R et al. Chaperonin-mediated protein folding at the surface of groEL through a 'molten globule'-like intermediate. Nature 1991; 352:36-42.
17. Buchner J, Schmidt M, Fuchs M et al. GroE facilitates refolding of citrate synthase by suppressing aggregation. Biochemistry 1991; 30:1586-1591.
18. Mendoza JA, Rogers E, Lorimer GH et al. Chaperonins facilitate the in vitro folding of monomeric mitochondrial rhodanese. Biol Chem 1991; 266:13044-1309.

19. Kubo T, Mizobata T, Kawata Y. Refolding of yeast enolase in the presence of the chaperonin GroE: The nucleotide specificity of GroE and the role of GroES. J Biol Chem 1993; 268:19346-19351.
20. Fisher MT. The effect of groES on the groEL-dependent assembly of dodecameric glutamine synthetase in the presence of ATP and ADP. J Biol Chem 1994; 269:13629-13636.
21. Schmidt M, Buchner J, Todd MJ et al. On the role of groES in the chaperonin-assisted folding reaction. Three case studies. J Biol Chem 1994; 269:10304-10311.
22. Hendrix RW. Purification and properties of groE, a host protein involved in bacteriophage assembly. J Mol Biol 1979; 129:375-392.
23. Hohn T, Hohn B, Engel A et al. Isolation and characterization of the host protein groE involved in bacteriophage lambda assembly. J Mol Biol 1979; 129:359-373.
24. Saibil H, Dong Z, Wood S et al. Binding of chaperonins. Nature 1991; 353:25-26.
25. Chen S, Roseman AM, Hunter AS et al. Location of a folding protein and shape changes in GroEL/GroES complexes imaged by cryo-electron microscopy. Nature 1994; 371:261-264.
26. Roseman AM, Chen S, White H et al. The chaperonin ATPase cycle: Mechanism of allosteric switching and movements of substrate-binding domains in GroEL. Cell 1996; 87:241-51.
27. Roseman AM, Ranson NA, Gowen B et al. Structures of unliganded and ATP-bound states of the Escherichia coli chaperonin GroEL by cryoelectron microscopy. J Struct Biol 2001; 135:115-25.
28. Ranson NA, Farr GW, Roseman AM et al. ATP-bound states of GroEL captured by cryo-electron microscopy. Cell 2001; 107:869-879.
29. Braig K, Otwinowski Z, Hegde R et al. The crystal structure of the bacterial chaperonin GroEL at 2.8 Å. Nature 1994; 371:578-586.
30. Boisvert DC, Wang J, Otwinowski Z et al. The 2.4 Å crystal structure of the bacterial chaperonin GroEL complexed with ATP-γ-S. Nat Struct Biol 1996; 3:170-7.
31. Hunt JF, Weaver AJ, Landry SJ et al. The crystal structure of the GroES co-chaperonin at 2.8 Å resolution. Nature 1996; 379:37-45.
32. Xu Z, Horwich AL, Sigler PB. The crystal structure of the asymmetric GroEL-GroES-(ADP)7 chaperonin complex. Nature 1997; 388:741-50.
33. Fiaux J, Bertelsen EB, Horwich AL et al. NMR analysis of a 900K GroEL GroES complex. Nature 2002; 418:207-11.
34. Fenton WA, Kashi Y, Furtak K et al. Residues in chaperonin GroEL required for polypeptide binding and release. Nature 1994; 371:614-619.
35. Landry SJ, Zeilstra-Ryalls J, Fayet O et al. Characterization of a functionally important mobile domain of GroES. Nature 1993; 364:255-8.
36. Shewmaker F, Maskos K, Simmerling C et al. The disordered mobile loop of GroES folds into a defined beta-hairpin upon binding GroEL. J Biol Chem 2001; 276:31257-64.
37. Kovalenko O, Yifrach O, Horovitz A. Residue lysine-34 in GroES modulates allosteric transitions in GroEL. Biochemistry 1994; 33:14974-8.
38. Torok Z, Vigh L, Goloubinoff P. Fluorescence detection of symmetric GroEL14(GroES7)2 hetero-oligomers involved in protein release during the chaperonin cycle. J Biol Chem 1996; 271:16180-16186.
39. Azem A, Diamant S, Kessel M et al. The protein-folding activity of chaperonins correlates with the symmetric GroEL14(GroES7)2 hetero-oligomer. Proc Natl Acad Sci USA 1995; 92:12021-12025.
40. Diamant S, Azem A, Weiss C et al. Increased efficiency of GroE-assisted protein folding by manganese ions. J Biol Chem 1995; 270:28387-28391.
41. Mayhew M, da Silva AC, Martin J et al. Protein folding in the central cavity of the GroEL-GroES chaperonin complex. Nature 1996; 379:420-426.
42. Brinker A, Pfeifer G, Kerner MJ et al. Dual function of protein confinement in chaperonin-assisted protein folding. Cell 2001; 107:223-233.
43. Ranson NA, White HE, Saibil HR. Chaperonins. Biochem J 1998; 333:233-242.
44. Richardson A, Landry SJ, Georgopoulos C. The ins and outs of a molecular chaperone machine. Trends Biochem Sci 1998; 23:138-43.
45. Sigler PB, Xu Z, Rye HS et al. Structure and function in GroEL-mediated protein folding. Annu Rev Biochem 1998; 67:581-608.
46. Wang JD, Weissman JS. Thinking outside the box: New insights into the mechanism of GroEL-mediated protein folding. Nat Struct Biol 1999; 6:597-600.
47. Weissman JS, Rye HS, Fenton WA et al. Characterisation of the active intermediate of a GroEL-GroES-mediated protein folding reaction. Cell 1996; 84:481-490.

48. Weissman JS, Hohl CM, Kovalenko O et al. Mechanism of GroEL action: Productive release of polypeptide from a sequestered position under GroES. Cell 1995; 83:577-587.
49. Rye HS, Burston SG, Fenton WA et al. Distinct actions of cis and trans ATP within the double ring of the chaperonin GroEL. Nature 1997; 388:792-798.
50. Rye HS, Roseman AM, Chen S et al. GroEL-GroES cycling: ATP and nonnative polypeptide direct alternation of folding-active rings. Cell 1999; 97:325-338.
51. Yifrach O, Horovitz A. Nested cooperativity in the ATPase activity of the oligomeric chaperonin GroEL. Biochemistry 1995; 34:5303-5308.
52. Lorimer G. Protein folding. Folding with a two-stroke motor. Nature 1997; 388:723.
53. Houry WA, Frishman D, Eckerskorn C et al. Identification of in vivo substrates of the chaperonin GroEL. Nature 1999; 402:147-154.
54. Wang JD, Herman C, Tipton KA et al. Directed evolution of substrate-optimized GroEL/S chaperonins. Cell 2002; 111:1027-1039.
55. Todd MJ, Viitanen PV, Lorimer GH. Dynamics of the chaperonin ATPase cycle: Implications for facilitated protein folding. Science 1994; 265:659-66.
56. Horovitz A, Fridmann Y, Kafri G et al. Allostery in chaperonins. J Struct Biol 2001; 135:104-114.
57. Viitanen PV, Gatenby AA, Lorimer GH. Purified chaperonin 60 (groEL) interacts with the nonnative states of a multitude of Escherichia coli proteins. Protein Sci 1992; 1:363-9.
58. Gordon CL, Sather SK, Casjens S et al. Selective in vivo rescue by GroEL/ES of thermolabile folding intermediates to phage P22 structural proteins. J Biol Chem 1994; 269:27941-51.
59. Brunschier R, Danner M, Seckler R. Interactions of phage P22 tailspike protein with GroE molecular chaperones during refolding in vitro. J Biol Chem 1993; 268:2767-72.
60. Ewalt KL, Hendrick JP, Houry WA et al. In vivo observation of polypeptide flux through the bacterial chaperonin system. Cell 1997; 90:491-500.
61. Chaudhuri TK, Farr GW, Fenton WA et al. GroEL/GroES-mediated folding of a protein too large to be encapsulated. Cell 2001; 107:235-46.
62. Farr GW, Fenton WA, Chaudhuri TK et al. Folding with and without encapsulation by cis- and trans-only GroEL-GroES complexes. EMBO J 2003; 22:3220-30.
63. Laemmli UK, Geguin F, Gujer-Kellenberger G. A factor preventing the major head protein of bacteriophage T4 from random aggregation. J Mol Biol 1970; 47:69-85.
64. Georgopoulos CP, Hendrix RW, Kaiser AD et al. Role of the host cell in bacteriophage morphogenesis: Effects of a bacterial mutation on T4 head assembly. Nature New Biol 1972; 239:38-41.
65. Takano T, Kakefuda T. Involvement of a bacterial factor in morphogenesis of bacteriophage capsid. Nature New Biol 1972; 239:34-37.
66. van der Vies SM, Gatenby AA et al. Bacteriophage T4 encodes a co-chaperonin that can substitute for Escherichia coli GroES in protein folding. Nature 1994; 368:654-656.
67. Keppel F, Rychner M, Georgopoulos C. Bacteriophage-encoded co-chaperonins can substitute for Escherichia coli's essential GroES protein. EMBO reports 2002; 3:893-898.
68. Hunt JF, van der Vies SM, Henry L et al. Structural adaptations in the specialized bacteriophage T4 co-chaperonin Gp31 expand the size of the Anfinsen cage. Cell 1997; 90:361-371.
69. Bakkes PJ, Faber BW, van Heerikhuizen H et al. The T4-encoded co-chaperonin, gp31, has unique properties that explain its requirement for the folding of the T4 major capsid protein. Proc Natl Acad Sci USA 2005; 102:8144-8149.
70. Viitanen PV, Lorimer GH, Seetheram R et al. Mammalian mitochondrial chaperonin 60 functions as a single toroidal ring. J Biol Chem 1992; 267:695-698.
71. Nielsen KL, Cowan NJ. A single ring is sufficient for productive chaperonin-mediated folding in vivo. Mol Cell 1998; 2:93-100.
72. Levy-Rimler G, Viitanen P, Weiss C et al. The effect of nucleotides and mitochondrial chaperonin 10 on the structure and chaperone activity of mitochondrial chaperonin 60. Europ J Biochem 2001; 268:3465-3472.
73. Richardson A, Schwager F, Landry SJ et al. The importance of a mobile loop in regulating chaperonin/ co-chaperonin interaction: Humans versus Escherichia coli. J Biol Chem 2001; 276:4981-4987.
74. Shewmaker F, Kerner MJ, Hayer-Hartl M et al. A mobile loop order-disorder transition modulates the speed of chaperonin cycling. Protein Sci 2004; 13:2139-2148.
75. Cavanagh AC, Morton H. The purification of early-pregnancy factor to homogeneity from human platelets and identification as chaperonin 10. Eur J Biochem 1994; 222:551-60.
76. Sayle R, Milner-White EJ. RasMol: Biomolecular graphics for all. Trends Biochem Sci 1995; 20:374.

CHAPTER 8

Co-Chaperones of the Endoplasmic Reticulum

Johanna Dudek, Martin Jung, Andreas Weitzmann, Markus Greiner and Richard Zimmermann*

Abstract

The rough endoplasmic reticulum (ER) plays a central role in the biogenesis of most extracellular and many organellar proteins in eukaryotic cells. Therefore, this organelle comprises molecular chaperones that are involved in import, folding/assembly, export, and degradation of polypeptides in millimolar concentrations. In addition, there are signal transduction components present in the ER membrane that affect and are affected by these processes at different levels. The ER lumenal Hsp70, termed BiP in mammals, is the central player in all these activities and involves a number of co-chaperones, i.e., ER-membrane integrated as well as ER-lumenal Hsp40s plus nucleotide exchange factors.

Introduction

The decisive initial step in the biogenesis of most extracellular and many organellar proteins of eukaryotic cells is their transport into the rough endoplasmic reticulum (ER) (Fig. 1A).[1] The same is true for biogenesis of many membrane proteins of various organelles (ER, Golgi apparatus, lysosomes, nucleus, peroxisomes) and the plasma membrane. Protein import into the ER can occur co or post-translationally and involves amino terminal signal peptides in the precursor proteins and a transport machinery. This machinery includes cytosolic components as well as components of both the ER-membrane and the ER-lumen.[2] Membrane insertion of precursor polypeptides as well as completion of their translocation are mediated by the ER protein translocase that comprises the Sec61p complex as central component. In addition, protein transport into the ER involves Hsp70-type molecular chaperones and their Hsp40-type co-chaperones plus nucleotide exchange factors (NEF). Typically, protein import into the ER is followed by folding or assembly of the transport substrates (Fig. 1B). Folding and assembly may involve molecular chaperones and folding catalysts.[3] Subsequently, the native proteins are delivered to their functional location by vesicular transport. In the case of mis-folding or mis-assembly the polypeptides are exported to the cytosol and degraded by the proteasome (Fig. 1C). Protein export involves a number of components that also play a role in protein import (such as Sec61p complex) as well as additional components that are indirectly involved in degradation (such as the ubiquitin-conjugating enzymes).[3-5] Hsp70-type molecular chaperones of the ER lumen (BiP in mammals, Kar2p in yeast) and their Hsp40-type co-chaperones (Sec63p, Scj1p, Jem1p in yeast) are also involved in export of mis-folded polypeptides to the cytosol for degradation by the proteasome (termed ERAD or ER associated degradation).[6-8]

*Corresponding Author: Richard Zimmermann—Medizinische Biochemie, Universität des Saarlandes, D-66421 Homburg, Germany. Email: bcrzim@med-rz.uni-saarland.de

Networking of Chaperones by Co-Chaperones, edited by Gregory L. Blatch.

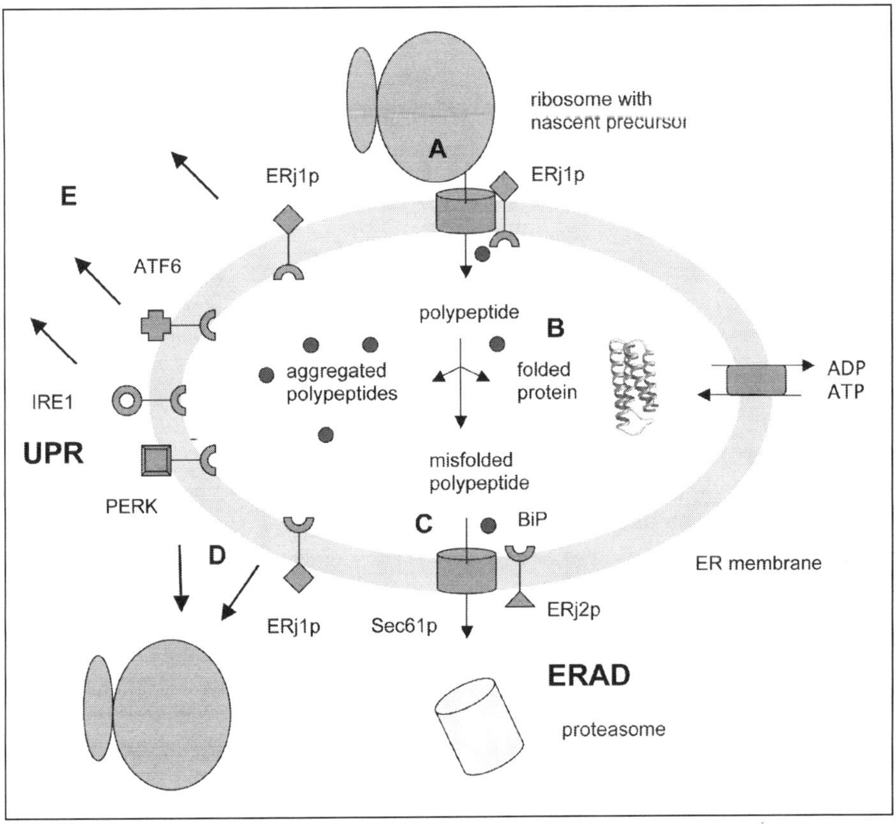

Figure 1. The mammalian rough endoplasmic reticulum (ER). The ER lumenal Hsp70 BiP is a central player in protein biogenesis in the ER that is involved in protein import (A), protein folding/assembly (B), protein export and protein degradation (C). It is part of a network of chaperones and co-chaperones in order to fulfill these functions (see Fig. 3A). In addition, BiP is part of a second network of interactions that has its role in signal transduction to the ribosomes (D) and the nucleus (E). A color version of this figure is available online at www.eurekah.com.

There are various signaling pathways that respond to protein mis-folding and mis-assembly (caused by stress or drugs such as tunicamycin and thapsigargin) and affect either translation or transcription (Fig. 1D,E), i.e., signal transduction to ribosomes and the genome, respectively, when ERAD is overwhelmed by mis-folded and aggregated polypeptides (termed UPR or unfolded protein response).[9,10] In mammals, this last activity involves the three ER membrane proteins PERK, ATF6 and IRE1, respectively (in yeast Ire1p). These proteins comprise lumenal domains (that are not related to J-domains) that interact with BiP and cytosolic domains that attenuate global translation (PERK) or induce selective transcription (ATF6, IRE1) in the absence of BiP.

To conceptualize the productivity of ER chaperone action, it is worth considering a major human secretory organ, e.g., the human pancreas. An average human being produces about 700 ml of pancreatic juice per day. The protein concentration of this body fluid is about 700 mg/ml, thus the daily production of secretory proteins in the human pancreas amounts to about 5 g. These 5 g correspond to about 100 µmol or 60 x 10^{18} molecules of secretory proteins per day.

Table 1. Chaperones and folding catalysts of the mammalian (A) or yeast (B) rough endoplasmic reticulum

	Chaperone Types	Co-Chaperone Types	Folding Catalysts	Signal Transduction Components (Affecting)
A	Hsp70s	Hsp40s NEF[1]	protein disulfide isomerases	IRE1[1] (transcription) ATF6[1] (transcription)
	Hsp90		peptidylprolyl-cis/trans-isomerases cyclophilin B FK506-binding protein 13	PERK[1] (translation) ERj1p (transcription+ translation)
	Calnexin Calreticulin			
B	Hsp70s	Hsp40s NEF[1]	protein disulfide isomerases	Ire1p[1]
	Calnexin		peptidylprolyl-cis/trans-isomerases cyclophilin 2 FK506-binding protein 2	

[1]The abbreviations used are: NEF: nucleotide exchange factor; IRE1: inositol-requiring enzyme; ATF6: activating transcription factor; PERK: double-stranded RNA-activated protein kinase-like ER kinase.

The Chaperone Network of the ER

Both the yeast and the mammalian ER contain molecular chaperones and folding catalysts in high concentrations.[3] On the one hand, these molecular chaperones belong to the classical Hsp40, Hsp70, and Hsp90 protein families (Table 1). However, a resident ER Hsp90, termed Grp94 in mammals, is absent from the yeast ER. On the other hand, the ER contains a special class of molecular chaperones (also termed lectins) that appear to be dedicated to the folding of glycoproteins.[11,12] The mammalian ER contains a soluble (calreticulin) as well as a membrane integrated (calnexin) lectin, the yeast ER just the membrane integrated one. The folding catalysts of the ER deal with either the formation of disulfide bonds (protein disulfide isomerases, PDI) or the isomerization of proline-containing peptide bonds (peptidyprolyl-cis/trans-isomerases, PPIase). The PPIases belong to either the cyclosporin A- or the FK506-sensitive protein family (cylophilin or FK506-binding protein).[13] Its noteworthy that the ER of a single cell type, typically, contains just a single type of cyclophilin and FK506-binding protein but may simultaneously contain several different PDIs. All these chaperones and folding catalysts have been observed to be present in larger complexes in various combinations (see below).

The Hsp70/Hsp40 Network of the ER

Just like the bacterial cytosol or the mitochondrial matrix, the ER contains the typical Hsp70 triad, comprising the Hsp70 itself (BiP, Kar2p) as well as a Hsp40-type co-chaperone and a nucleotide exchange factor (NEF). These proteins have also been shown to be able to perform the classical Hsp70 reaction cycle (Fig. 2), thereby mediating the folding and/or assembly of a newly-synthesized as well as newly–imported polypeptide. Similarly to the two above-mentioned cellular compartments, there are two Hsp70-type chaperones in both the yeast as well as the mammalian ER (Table 2).[14-29] One of them, however, may also be referred to as a Hsp110 protein family member (Grp170, Lhs1p).[14,22,30] The Hsp40-type co-chaperones can be divided into membrane proteins with a lumenal J-domain and into soluble lumenal proteins (Table 2, Fig. 3). Furthermore, they can be classified according to the domains they have in common with the bacterial DnaJ protein (i.e., besides the actual J-domain).[7,13,31-47] Thus only the ER-lumenal Hsp40s most closely related to DnaJ (Scj1p, ERj3p, and ERj4p)

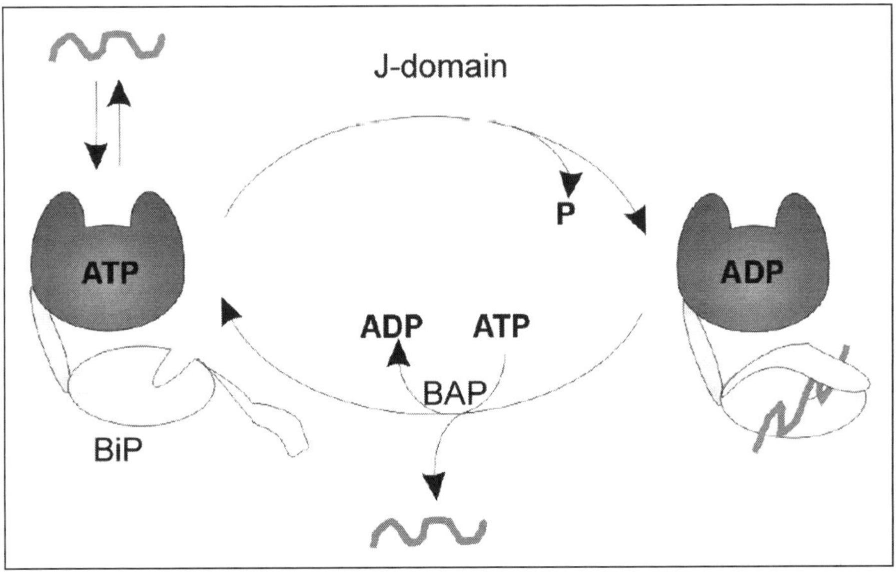

Figure 2. The BiP functional cycle. In the ATP-bound state, BiP has a low affinity for substrate polypeptides. Upon binding of a J-domain to the underside of the ATP-binding cleft of BiP, ATP is hydrolyzed and a conformational change is triggered in the peptide binding domain. As a result, any polypeptide substrate that bound to the peptide binding domain of BiP either spontaneously or as delivered by an ERjp becomes trapped in the peptide binding domain. Nucleotide exchange that may or may not have been stimulated by a nucleotide exchange factor, such as BAP, reverses the conformational shift and allows the polypeptide substrate to leave the peptide binding domain. Subsequently, the polypeptide substrate may fold or enter another round of binding and release. Actually, we are convinced that a transient substrate-like interaction of the J-domain with the peptide binding domain of BiP may be part of the reaction cycle.[62] A color version of this figure is available online at www.eurekah.com.

can be expected to have the ability to bind substrate polypeptides and to deliver them to BiP, i.e., to facilitate a Hsp70 reaction cycle in analogy to DnaJ in *E. coli.*[13,34,35,43,44] Thus ERj3p plus ERj4p may be viewed as a pair of soluble plus membrane-integrated Hsp40s in the mammalian ER, in analogy to the above mentioned pair of lectins (calreticulin plus calnexin). The current state of knowledge in this area of research is most complicated with respect to the nucleotide exchange factors. There seems to be a bona fide functional equivalent to bacterial GrpE in the ER lumen (BAP, Sil1p; see Chapter by Brodsky and Bracher).[48,49] Furthermore, this ER-lumenal NEF appears to be closely related to HspBP1, the NEF of cytosolic Hsp70 in eukaryotes.[50] In addition, in yeast Lhs1p has been shown to be able to act as a NEF for Kar2p (see below).[48,51] The fact that the simultaneous deletion of Lhs1p and Sil1p results in synthetic lethality further supports the notion that these two proteins provide functional redundancy. Various members of the resident ER Hsp70-cycle have been found in large complexes with each other, with other chaperones and folding catalysts, and with other resident ER proteins that are involved in N- or O-glycosylation (UDP-glucose-glycoprotein-glycosyltransferase, SDF2L1p) and calcium homeostasis (calumenin, reticulocalbin) (Table 2).[52-54]

From the structural point of view, the most interesting Hsp40 in the lumen of the mammalian ER is ERj5p since it comprises a J- as well as four thioredoxin domains and thus can be expected to provide a direct link between polypeptide folding and disulfide bridge formation.[38,41] It may be able to recruit BiP to sites that require disulfide bond reshuffling. Alternatively, the thioredoxin domains may provide a mechanism for redox-regulation of the J-domain.

Table 2. Hsp70-chaperones of the mammalian (A) or yeast (B) rough endoplasmic reticulum and their co-chaperones

A	Protein Families	J Protein Type[61]	Stress Induced	Yeast Ortholog	Stably Interacting Proteins/Complexes
BiP (Grp78)	DnaK/Hsp70		+	Kar2p	Grp94, PDIs, cyclophilin B, calnexin, calreticulin, Grp170, ERj3p, UDP-glc-gp-GT, SDF2L1p[1]
Grp170	DnaK/Hsp70		+	Lhs1p	Grp94, PDIs, cyclophilin B, calnexin, calreticulin BiP, ERj3p, UDP-glc-gp-GT, SDF2L1p[1]
ERj1p (Mtj1p)	DnaJ/Hsp40	III	-	none	BiP, ribosomes
ERj2p (Sec63p)	DnaJ/Hsp40	III	-	Sec63p	BiP, Sec61p, Sec62p, calumenin, reticulocalbin
ERj3p (HEDJ, Dj9)	DnaJ/Hsp40	II	+	Scj1p	BiP, SDF2L1p[1]
ERj4p (ERdj4p, MDG1)	DnaJ/Hsp40	II	+	none	BiP
ERj5p (ERdj5p, JPDI)	DnaJ/Hsp40	III	+	none	BiP
BAP	NEF[1]		-	Sil1p	

B	Protein Families	J Protein Type[61]	Stress Induced	Mammalian Ortholog	Essential Gene	Interacting Proteins/Complexes
Kar2p	DnaK/Hsp70		+	BiP	yes	Sec63p, Lhs1p, Sil1p
Lhs1p (Ssi1p, Cer1p)	DnaK/Hsp70		+	Grp170	no/synthetic-lethality	Kar2p
Scj1p	DnaJ/Hsp40	I	+	ERj3p	no	
Scj2p (YFR041C)	DnaJ/Hsp40	III	+	none	not determined	
Sec63p (Ptl1p, Npl1p)	DnaJ/Hsp40	III	-	ERj2p	yes	Sec61p, Sbh1p, Sss1p, Sec62p, Sec71p/72p, Kar2p
Jem1p	DnaJ/Hsp40	III	+	none	no	
Sil1p	NEF[1]		+	BAP	no/synthetic-lethality	

[1]The abbreviations used are: NEF: nucleotide exchange factor; SDF2L1p: stromal cell-derived factor 2 like protein 1.

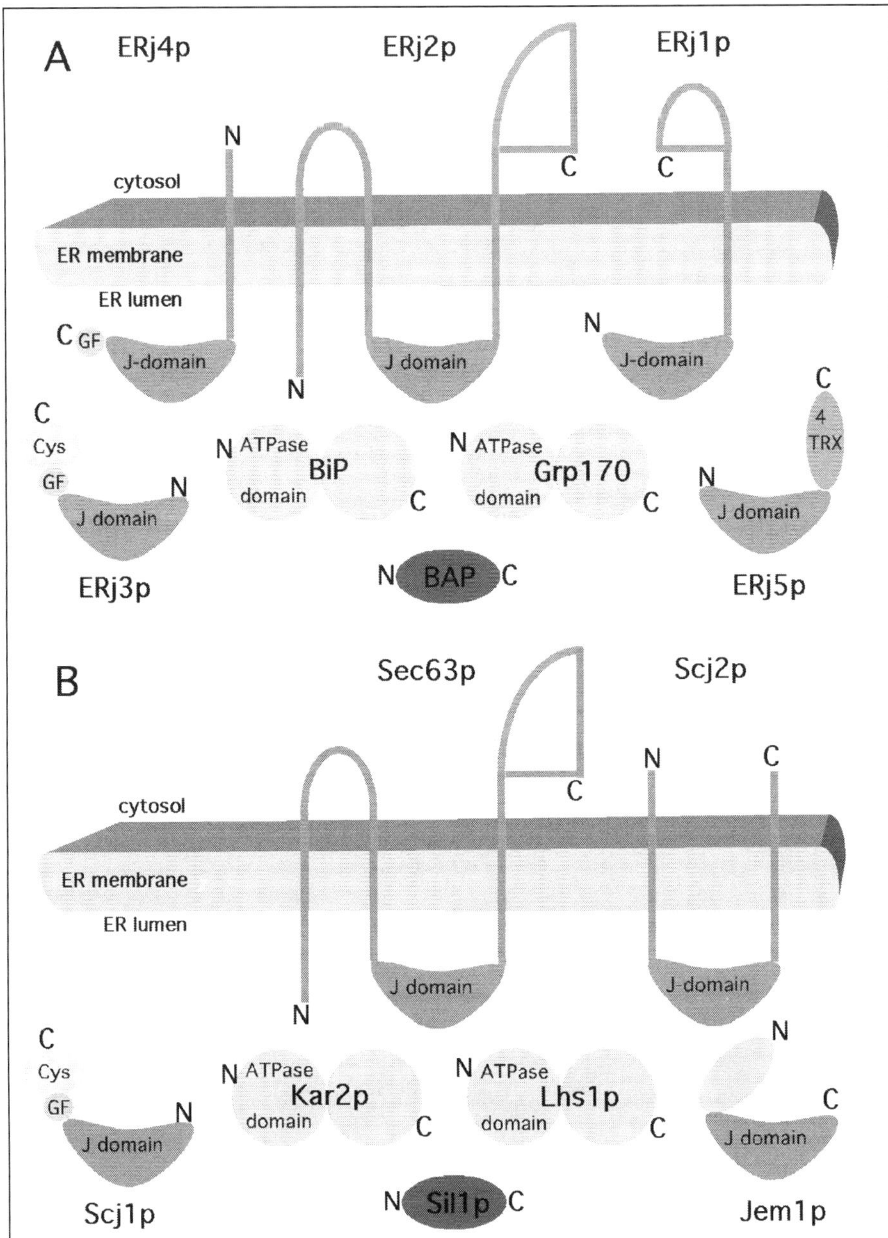

Figure 3. The network of Hsp70-chaperones and co-chaperones that is present in the mammalian- (A) and yeast- (*Saccharomyces cerevisiae*; B) rough endoplasmic reticulum. N: amino terminus; C: carboxy terminus; G/F: Gly/Phe-rich region; Cys: Cys-rich region (we note that this region represents the four zinc-finger-like motif in Scj1p but not in ERj3p and that, in contrast to DnaJ, the four zinc-finger-like motif in Scj1p seems to be organized through disulfide bonds rather than the coordination of metal ions); TRX: thioredoxin domain. A color version of this figure is available online at www.eurekah.com.

The Role of Co-Chaperones of BiP/Kar2p in Protein Transport into the ER

Hsp70-type molecular chaperones of the ER lumen, i.e., BiP plus Grp170 in mammals or Kar2p plus Lhs1p in yeast, are involved in co- and post-translational insertion of precursor polypeptides into the Sec61p complex.[14,26,29] BiP/Kar2p was also identified as a lumenal protein that is involved in the completion of protein translocation.[20,26,29] Furthermore, BiP was shown to seal the lumenal end of the mammalian Sec61p complex in the absence of protein translocation and at several stages during co-translational translocation of preproteins.[16-19,21] Currently, the central questions are how these various activities of BiP/Kar2p can be integrated into a single model and how co-chaperones come into play.

Electron microscopic as well as electrophysiological analyses are consistent with the idea that the active protein translocase in the yeast and mammalian microsomal membrane comprises as core component an oligomer that contains four heterotrimeric Sec61p subcomplexes (Fig. 4).[16-19,21,55-58] We propose that the oligomeric Sec61p complex involves a reaction cycle that includes an inactive ion channel state that is activated by precursor polypeptides (i.e., substrate) in two steps and is directly linked to the reaction cycle of BiP at two different stages.[21] We propose that upon contact of a precursor polypeptide with a single Sec61p subcomplex on the cis-side of the ER-membrane there is a conformational change in this subcomplex and, as a result, a path for ions and small molecules is opened at the level of this heterotrimeric subcomplex (first step of activation). This leads to allosteric activation of the neighboring subcomplexes and, subsequently, to formation of a large path for both ions and the polypeptide substrate that is formed at the subunit interface (second step of activation). Eventually, the precursor polypeptide is released on the trans side of the ER-membrane and the oligomeric complex returns stepwise to the inactive ion channel state (same steps as in activation but reversed order). Furthermore, we envision the role of BiP in insertion of precursor polypeptides into the Sec61p complex as well as in sealing the lumenal end of the Sec61p complex as related. In both cases, BiP acts on the conformation of a subunit of the Sec61p complex and, thereby, facilitates formation and disintegration of the large channel, respectively. In addition, BiP works as a molecular ratchet on the precursor polypeptide chain as soon as it has access to it on the trans side of the membrane.[20] Therefore, at this stage BiP can be replaced by artificial polypeptide chain binding proteins, such as antibodies or avidin (i.e., in the case of biotinylated precursors). We assume that the role of the second Hsp70-type chaperone (Lhs1p and Grp170) in some or all these activities is in nucleotide exchange on BiP.

There is no doubt that the physical and mechanistic link between the Sec61p- and the BiP-reaction cycle is provided by a membrane integrated Hsp40 with a lumenal J-domain. In yeast, Sec63p has been shown to provide the lumenal J-domain that allows Kar2p to play its role in insertion of precursors into the Sec61p complex as well as in completion of translocation.[26-29] Furthermore, it has been shown that human ERj1p (Htj1p) can complement the otherwise lethal deletion of Sec63p in yeast.[47] For the mammalian system it was observed that a resident ER protein with a lumenal J-domain is involved in sealing of the Sec61p complex.[18] In addition, in pancreatic microsomes ERj2p (Sec63p) was found in association with the Sec61p complex and to be present in approximately stoichiometric amounts as compared to heterotrimeric Sec61p complexes.[45] Furthermore, ERj1p that does not have an ortholog in yeast was observed in association with translating ribosomes in the same microsomes.[36,39,40] Therefore, we propose that in the mammalian ER two different membrane proteins provide J-domains in the neighborhood of translating ribosomes and Sec61p complexes and allow BiP to play its various roles in protein import, ERj1p and ERj2p (Fig. 1). At present, however, it is not clear what the specialization of the two Hsp40s may be (e.g., co- versus post-translational protein import; protein import versus protein export). In any case, the dual role of the respective J-domain must reside in recruiting BiP/Kar2p to the Sec61p complex and the transport substrate, respectively, and in simultaneously stimulating the ATPase activity in order to allow substrate binding.

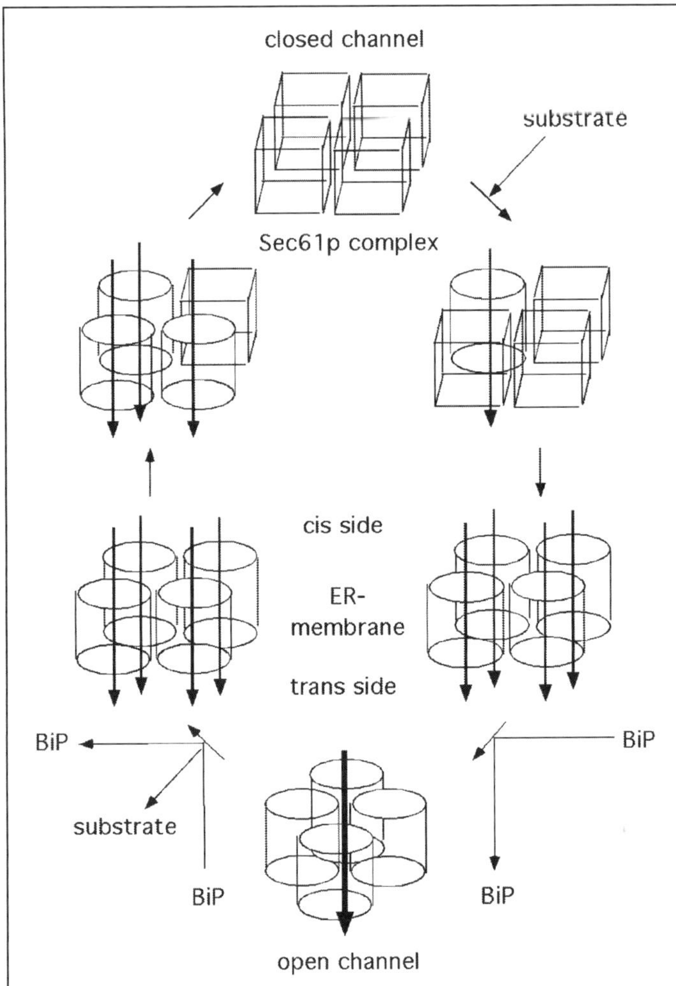

Figure 4. The reaction cycle of the Sec61p complex is coupled to the BiP reaction cycle. The Sec61p complex is shown as a tetramer without its surrounding membrane (at two stages of the cycle the plane of the membrane as well as the cytosolic/cis and lumenal/trans faces of the membrane are indicated). The path for ions at the level of the Sec61p subcomplexes is indicated by a thin arrow, the path for both ions and polypeptide substrate at the subunit interface is represented by the fat arrow. See text for details.

ERj1p is special for other reasons, too.[36,40,59] It appears to have regulatory roles that are related to transcription as well as to translation, i.e., in addition to its above-mentioned role in protein import (Fig. 1). The cytosolic domain of ERj1p has the ability to allosterically inhibit translation at the stage of initiation when it is not bound to BiP. Furthermore, ERj1p has all the features of a membrane-tethered transcription factor that can be activated by regulated intra-membrane proteolysis (termed RIP). The cytosolic domain has actually been shown to be able to enter the nucleus. Thus, ERj1p appears to be the central player in protein biogenesis at the ER-surface, it combines in a single molecule the BiP-recruiting abilities of Sec63p and the various facets of regulation of gene expression that are also covered by PERK, ATF6 and IRE1.

Open Questions Related to the Networking of ER Chaperones by Co-Chaperones

Although we have captured the first snap shots of the action of the various co-chaperones of ER resident Hsp70-chaperones, we are still far from a complete picture of this dynamic chaperone network and its regulatory mechanisms. In mammalian organisms, it has yet to be demonstrated that the various co-chaperones of BiP are simultaneously present in the ER of a single cell. The quantitative aspects have hardly been considered yet. Thus in the future, we need to address the concentrations of the various members of the network in the ER of intact cells and the affinities of the various proteins or domains for each other. The putative chaperone functions of Grp170 and Lhs1p have to be studied in further detail. The interaction of Lhs1p with Kar2p has to be characterized at the structural level. Furthermore, the specializations of the various Hsp40s will have to be analyzed in greater detail.

In the eukaryotic cytosol the Hsp70 cycle is linked to the Hsp90 cycle. Thus one wonders to what extent this is true for the ER. Furthermore, why does the mammalian ER contain Grp94, while the yeast ER does not. One possibility is that this chaperone has its main role in the loading of MHC class I molecules with peptides for subsequent presentation on the cell surface, rather than the folding or assembly of polypeptides.

One of the central puzzles is related to the fact that BiP as well as Grp170 and Grp94 depend on the hydrolysis of ATP for their activity. However, the mechanism of ATP-transport into the ER has remained elusive. The most likely scenario is the presence of an ADP/ATP-carrier in the ER-membrane that allows exchange of ADP versus ATP (Fig. 1).

Other open questions are related to the role of various ER-chaperones and folding catalysts in calcium homeostasis, or vice versa, or both. The ER plays a central role in calcium homeostasis. Furthermore, it's a fact that many of the chaperones and folding catalysts are calcium-binding proteins (therefore they were also termed CaBPs).[60] BiP has been shown to be affected functionally by the concentration of calcium and, in general, ER-calcium depletion (e.g., by drugs such as thapsigargin) triggers mis-folding of polypeptides and, therefore, the UPR. However, the exact inter-relationship between the two remains to be resolved, since high calcium concentrations inhibit BiP's ATPase activity. Furthermore, ERj2p has been found to be associated with the CaBPs calreticulin, calumenin and reticulocalbin. This may be related to the ion channel activity of the activated Sec61p complex.

Acknowledgements

The authors work on this subject was supported by the Deutsche Forschungsgemeinschaft.

References

1. Palade G. Intracellular aspects of the process of protein synthesis. Science 1975; 189:347-358.
2. Jung M, Bies C, Dudek J et al. Multiple roles of molecular chaperones in protein transport into the mammalian endoplasmic reticulum. Research Signpost, recent research developments in Molecular & Cellular Biology 2003; 3:549-562.
3. Frien M, Jung M, Völzing C et al. The rough endoplasmic reticulum: A proteomic analysis. Research Signpost, recent research developments in Biochemistry 2003; 4:113-124.
4. Wiertz EJH, Tortoralla D, Bogyo M et al. Sec61-mediated transfer of a membrane protein from the endoplasmic reticulum to the proteasome for destruction. Nature 1996; 384:432-438.
5. Pilon M, Schekman R, Römisch K. Sec61p mediates export of a misfolded secretory protein from the endoplasmic reticulum to the cytosol for degradation. EMBO J 1997; 16:4540-4548.
6. Plemper RK, Bömler S, Bordallo J et al. Mutant analysis links the translocon and BiP to retrograde protein transport for ER degradation. Nature 1997; 388:891-895.
7. Nishikawa S-I, Fewell SW, Kato Y et al. Molecular chaperones in the yeast endoplasmic reticulum maintain the solubility of proteins for retrotranslocation and degradation. J Cell Biol 2001; 153:1061-1069.
8. Knittler MR, Dirks S, Haas IG. Molecular chaperones involved in protein degradation in the endoplasmic reticulum: Quantitative interaction of the heat shock cognate BiP with partially folded immunoglobulin light chains that are degraded in the endoplasmic reticulum. Proc Natl Acad Sci USA 1995; 92:764-1768.

9. Ma Y, Hendershot LM. The unfolding tale of the unfolded protein response. Cell 2001; 107:827-830.
10. Zhang K, Kaufman RJ. Signaling the unfolded protein response from the endoplasmic reticulum. J Biol Chem 2004; 279:25935-25938.
11. Bergeron JJM, Brenner MB, Thomas DY et al. Calnexin: A membrane-bound chaperone of the endoplasmic reticulum. Trends Biochem Sci 1994; 19:124-128.
12. Bleackley RC, Atkinson EA, Burns K et al. Calreticulin: A granule-protein by default or design? Curr Top Microbiol Immunol 1995; 198:145-159.
13. Bies C, Guth S, Janoschek K et al. A Scj1p homolog and folding catalysts present in dog pancreas microsomes. Biol Chem 1999; 380:175-1182.
14. Dierks T, Volkmer J, Schlenstedt G et al. A microsomal ATP-binding protein involved in efficient protein transport into the mammalian endoplasmic reticulum. EMBO J 1996; 15:6931-6942.
15. Kassenbrock CK, Kelly RB. Interaction of heavychain binding protein (BiP/GRP78) with adenine nucleotides. EMBO J 1989; 8:1461-1467.
16. Hamman BD, Hendershot LM, Johnson AE. BiP maintains the permeability barrier of the ER membrane by sealing the lumenal end of the translocon pore before and early in translocation. Cell 1998; 92:747-758.
17. Haigh NG, Johnson AE. A new role for BiP: Closing the aqueous translocon pore during protein integration into the ER membrane. J Cell Biol 2002; 156:261-270.
18. Alder NN, Shen Y, Brodsky JL et al. The molecular mechanism underlying BiP-mediated gating of the Sec61 translocon of the endoplasmic reticulum. J Cell Biol 2005; 168:389-399.
19. Liao S, Lin J, Do H et al. Both lumenal and cytosolic gating of the aqueous ER translocon pore are regulated from inside the ribosome during membrane protein integration. Cell 1997; 90:31-42.
20. Tyedmers J, Lerner M, Wiedmann M et al. Polypeptide chain binding proteins mediate completion of cotranslational protein translocation into the mammalian endoplasmic reticulum. EMBO rep 2003; 4:505-510.
21. Wirth A, Jung M, Bies C et al. The Sec61p complex is a dynamic precursor activated channel. Mol Cell 2003; 12:261-268.
22. Craven RA, Egerton M, Stirling CJ. A novel Hsp70 of the yeast ER lumen is required for the efficient translocation of a number of protein precursors. EMBO J 1996; 15:2640-2650.
23. Sanders SL, Whitfield KM, Vogel JP et al. Sec61p and BiP directly facilitate polypeptide translocation into the ER. Cell 1992; 69:353-365.
24. Vogel JP, Misra LM, Rose MD. Loss of BiP/GRP78 function blocks translocation of secretory proteins in yeast. J Cell Biol 1990; 110:1885-1895.
25. Matlack KES, Misselwitz B, Plath K et al. BiP acts as a molecular ratchet during posttranslational transport of prepro-α factor across the ER membrane. Cell 1999; 97:553-564.
26. Young BP, Craven RA, Reid PJ et al. Sec63p and Kar2p are required for the translocation of SRP-dependent precursors into the yeast endoplasmic reticulum in vivo. EMBO J 2001; 20:262-271.
27. Lyman SK, Schekman R. Interaction between BiP and Sec63p is required for the completion of protein translocation into the ER of Saccharomyces cerevisiae. J Cell Biol 1995; 131:1163-1171.
28. Lyman SK, Schekman R. Binding of secretory precursor polypeptides to a translocon subcomplex is regulated by BiP. Cell 1997; 88:85-96.
29. Brodsky JL, Goeckeler J, Schekman R. BiP and Sec63p are required for both co and posttranslational protein translocation into the yeast endoplasmic reticulum. Proc Natl Acad Sci USA 1995; 92:9643-9646.
30. Easton DP, Kaneko Y, Subjeck JR. The Hsp110 and Grp170 stress proteins: Newly recognized relatives of the Hsp70s. Cell Stress & Chaperones 2000; 5:276-290.
31. Toyn J, Hibbs AR, Sanz P et al. In vivo and in vitro analysis of ptl1, a yeast ts mutant with a membrane-associated defect in protein translocation. EMBO J 1988; 7:4347-4353.
32. Rothblatt JA, Deshaies RJ, Sanders SL et al. Multiple genes are required for proper insertion of secretory proteins into the endoplasmic reticulum in yeast. J Cell Biol 1989; 109:2641-2652.
33. Sadler I, Chiang A, Kurihara T et al. A yeast gene important for protein assembly into the endoplasmic reticulum and the nucleus has homology to DnaJ, an Escherichia coli heat shock protein. J Cell Biol 1989; 109:2665-2675.
34. Schlenstedt G, Harris S, Risse B et al. A yeast DnaJ homologue, Scj1p, can function in the endoplasmic reticulum with BiP/Kar2p via a conserved domain that specifies interactions with Hsp70. J Cell Biol 1995; 129:979-988.
35. Bies C, Blum R, Dudek J et al. Characterization of pancreatic ERj3p, a homolog of yeast DnaJ-like protein Scj1p. Biol Chem 2004; 385:389-395.
36. Blau M, Mullapudi S, Becker T et al. ERj1p uses a universal ribosomal adaptor site to coordinate the 80S ribosome at the membrane. Nat Struct Mol Biol 2005; 12:1015-1016.

37. Brightman SE, Blatch GL, Zetter BR. Isolation of a mouse cDNA encoding MTJ1, a new murine member of the DnaJ family of proteins. Gene 1995; 153:249-254.
38. Cunnea PM, Miranda-Vizuete A, Bertoli G et al. ERdj5, an endoplasmic reticulu (ER)-resident protein containing DnaJ and thioredoxin domains, is expressed in secretory cells or following stress. J Biol Chem 2003; 278:1059-1066.
39. Dudek J, Volkmer J, Bies C et al. A novel type of co-chaperone mediates transmembrane recruitment of DnaK-like chaperones to ribosomes. EMBO J 2002; 21:2958-2967.
40. Dudek J, Greiner M, Müller A et al. ERj1p plays a basic role in protein biogenesis at the endoplasmic reticulum. Nat Struct Mol Biol 2005; 12:1008-1014.
41. Hosoda A, Kimata Y, Tsuru A et al. JPDI, a novel endoplasmic reticulum-resident protein containing both a BiP-interacting J-domain and thioredoxin-like motifs. J Biol Chem 2003; 278:2669-2676.
42. Meyer H-A, Grau H, Kraft R et al. Mammalian Sec61 is associated with Sec62 and Sec63. J Biol Chem 2000; 275:14550-14557.
43. Shen Y, Meunier L, Hendershot LM. Identification and characterization of a novel endoplasmic reticulum (ER) DnaJ homologue, which stimulates ATPase activity of BiP in vitro and is induced by ER stress. J Biol Chem 2002; 277:15947-15956.
44. Shen Y, Hendershot LM. ERdj3p, a stress-inducible endoplasmic reticulum DnaJ homologue, serves as a cofactor for BiP's interactions with unfolded substrates. Mol Biol Cell 2004; 16:40-50.
45. Tyedmers J, Lerner M, Bies C et al. Homologs of the yeast Sec complex subunits Sec62p and Sec63p are abundant proteins in dog pancreas microsomes. Proc Natl Acad Sci USA 2000; 97:7214-7219.
46. Yu M, Haslam RHA, Haslam DB. HEDJ, an Hsp40 co-chaperone localized to the endoplasmic reticulum of human cells. J Biol Chem 2000; 275:24984-24992.
47. Kroczynska B, Evangelista CM, Samant SS et al. The SANT2 domain of murine tumor cell DnaJ-like protein 1 human homologue interacts with α_1-antichymotrypsin and kinetically interferes with its serpin inhibitory activity. J Biol Chem 2004; 279:11432-11443.
48. Tyson JR, Stirling CJ. LHS1 and SIL1 provide a lumenal function that is essential for protein translocation into the endoplasmic reticulum. EMBO J 2000; 19:6440-6452.
49. Chung KT, Shen Y, Hendershot LM. BAP, a mammalian BiP associated protein, is a nucleotide exchange factor that regulates the ATPase activity of BiP. J Biol Chem 2002; 277:47557-47563.
50. Shomura Y, Dragovic Z, Chang H-C et al. Regulation of Hsp70 function by HspBP1: Structural analysis reveals an alternate mechanism for Hsp70 nucleotide exchange. Mol Cell 2005; 17:367-379.
51. Steel GJ, Fullerton DM, Tyson JR et al. Coordinated activation of Hsp70 chaperones. Science 2004; 303:98-101.
52. Tatu U, Helenius A. Interactions between newly synthesized glycoproteins, calnexin and a network of resident chaperones in the endoplasmic reticulum. J Cell Biol 1997; 136:555-565.
53. Meunier L, Usherwood Y-K, Chung KT et al. A subset of chaperones and folding enzymes form multiprotein complexes in endoplasmic reticulum to bind nascent proteins. Mol Biol Cell 2002; 13:4456-4469.
54. Tyedmers J, Lerner M, Nastainczyk W et al. Calumenin and reticulocalbin are associated with the protein translocase of the mammalian endoplasmic reticulum. J Biol Sci 2005; 5:70-75.
55. Beckmann R, Bubeck D, Grassucci R et al. Alignment of conduits for the nascent polypeptide chain in the ribosome-Sec61 complex. Science 1997; 278:2123-2126.
56. Beckmann R, Spahn CM, Eswar N et al. Architecture of the protein-conducting channel associated with the translating 80S ribosome. Cell 2001; 107:361-372.
57. Hanein D, Matlack KES, Jungnickel B et al. Oligomeric rings of the Sec61p complex induced by ligands required for protein translocation. Cell 1996; 87:721-732.
58. Menetret JF, Neuhof A, Morgan DG et al. The structure of ribosome-channel complexes engaged in protein translocation. Mol Cell 2000; 6:1219-1232.
59. Zupicich J, Brenner SE, Skarnes WC. Computational prediction of membrane-tethered transcription factors. Genome Biol 2001; 2:50.1-50.6.
60. Van PN, Peter F, Söling H-D. Four intracisternal calcium-binding glycoproteins from rat liver microsomes with high affinity for calcium. J Biol Chem 1989; 264:17494-17501.
61. Cheetham ME, Caplan AJ. Structure, function and evolution of DnaJ: Conservation and adaptation of chaperone function. Cell Stress Chap 1998; 3:28-36.
62. Hennessy F, Nicoll WS, Zimmermann R et al. Not all J domains are created equal: Implications for the specificity of Hsp40-Hsp70 interactions. Prot Sci 2005; 14:1697-1709.

CHAPTER 9

The Evolution and Function of Co-Chaperones in Mitochondria

Dejan Bursać and Trevor Lithgow*

Abstract

Mitochondrial chaperones mediate and affect critical organellar processes, essential for cellular function. These chaperone systems have both prokaryotic and eukaryotic features. While some of the mitochondrial co-chaperones have clear homologues in prokaryotes, some are unique to eukaryotes and have no homologues in the chaperone machinery of other cellular compartments. The mitochondrial co-chaperones are required for protein import into the organelle and in enforcing the structure of the main chaperones. In addition to novel types of interaction with their senior partners, unexpected and essential interactions between the co-chaperones themselves have recently been described.

Introduction

The mitochondrion is an endosymbiotically derived double membrane organelle of prokaryotic origin, characteristic of eukaryotic organisms. The organelle still retains many prokaryotic features, such as 70S ribosomes, the machinery to synthesize FeS clusters and its own circular chromosomal DNA. The mitochondrion is enveloped by two membranes, the outer and the inner membrane. All mitochondrial co-chaperones identified to date are localized in the lumen of the inner membrane, the mitochondrial matrix. While comparative analyses of mitochondrial chaperones with their bacterial homologues have yielded rich understanding of the processes of protein folding and FeS cluster assembly, the co-chaperones regulating protein import are unique to mitochondria and must have evolved some time after the original endosymbiont began to assume the form of an organelle. Much of the understanding of mitochondrial function comes from studies in yeast *Saccharomyces cerevisiae*, and their co-chaperone complement will be examined in this chapter.

The Mitochondrial Homologue of DnaK and Its Co-Chaperones

The major mitochondrial Hsp70 chaperone (called Ssc1 in yeast) is regulated by 5 co-chaperones of the J protein family. These co-chaperones define the spectrum of processes Ssc1 mediates, such as protein folding, import and degradation. Protein folding is a conserved process, mediated in the mitochondrial matrix in an analogous way to the process in the bacterial cytoplasm, and two co-chaperones regulating the process in mitochondria (Mdj1 and Mge1) are mitochondrial equivalents of the bacterial co-chaperones DnaJ and GrpE. Mdj1, a DnaJ

*Corresponding Author: Trevor Lithgow—Department of Biochemistry and Molecular Biology, Bio21 Institute of Molecular Science and Biotechnology, University of Melbourne, Parkville, Victoria 3010, Australia. Email: t.lithgow@unimelb.edu.au

Networking of Chaperones by Co-Chaperones, edited by Gregory L. Blatch.

homologue, is the only type I J protein in mitochondria.[1] Mge1, a GrpE homologue is the only bacterial-type nucleotide exchange factor in eukaryotes.[2]

The mitochondrial Ssc1/Mdj1/Mge1 chaperone system shares biochemical, as well as functional properties with the bacterial DnaK/DnaJ/GrpE (Fig. 1). Like DnaJ, Mdj1 binds to and prevents the aggregation of unfolded firefly luciferase in in vitro assays.[3] The unfolded substrate is delivered to the major chaperone and a typical J/Hsp70 interaction follows: Mdj1, through its J domain stimulates the ATPase activity of Ssc1, coupling it with substrate delivery.[4] Firefly luciferase folding is facilitated by Ssc1, but the substrate is efficiently released only in the presence of Mge1, which affects ADP release from the chaperone. Nucleotide release factors in other eukaryotic compartments are not of bacterial origin (see the Chapter in this volume by Brodsky and Bracher), and the presence of this bacterial-type co-chaperone further underlines the prokaryotic nature of the mitochondrial chaperone system.

Mdj1 can both deliver substrate to DnaK, and stimulate the ATPase activity of DnaK in vitro, resulting in a productive folding interaction.[5] If expressed in bacterial cells, Mge1 can replace GrpE. Also functional in bacterial cells is a hybrid J protein composed of the glycine-rich and zinc finger domains of DnaJ fused to the J domain of Mdj1.[6] Despite this apparent similarity between the co-chaperones of mitochondria and bacteria, the two chaperone machines are not completely equivalent. Ssc1 can not, even in the presence of Mdj1, complement the loss of DnaK in vivo, and full-length Mdj1 can not complement the loss of DnaJ.[6]

Folding of proteins within the mitochondrial matrix is a major function of Ssc1[7] and is regulated by Mdj1. Analysis of mutant forms of Mdj1 has revealed that folding of mitochondrial proteins following their import into the organelle largely depends on the co-chaperone, but that it only interacts with fully translocated substrates and does not play a role in protein import itself.[1,8] Increased levels of misfolded proteins and protein aggregates were observed in organello upon heat shock of mitochondria in yeast cells genetically depleted of Mdj1, confirming its importance in this process.[3] Protein folding in mitochondria is not solely dependent on Ssc1; some substrates are handed over to the conserved chaperonin system, Hsp60/Hsp10 in yeast (homologues of bacterial GroES and GroEL, respectively), which completes the folding reactions.[9,10]

Ssc1 also assists the folding of proteins translated from the mitochondrial 70S ribosomes. Mdj1 and Ssc1 were found associated with the nascent chain of Var1 emerging from the mitochondrial ribosomes and are proposed to protect the emerging protein from aggregation in the protein-dense mitochondrial matrix. In the absence of functional Mdj1, Var1 readily aggregates and does not assume its biologically active form, suggesting that the co-chaperone is also essential for its folding.[11]

Yeast cells genetically depleted of Mdj1 either completely (*rho*^{o}) or partially (*rho*$^-$) lose mitochondrial DNA, though the details of how this occurs remain to be understood. Partial loss of mtDNA seems likely due, at least in part, to inefficient assembly of fidelity regulating components with the DNA polymerase, and the polymerase itself, in the absence of the co-chaperone.[12] In bacteria DnaJ mediates disassembly of the DNA replication machinery during λ phage DNA replication, but Mdj1 has not been shown to exert such direct effects on the mtDNA replication components.

Mitochondrial Protein Import with a Highly Advanced Hsp70 Machine at Its Core

Protein translocation across membranes is an essential cellular process, in some compartments mediated by Hsp70/J protein partnerships.[13] Nearly all mitochondrial proteins are encoded by the nuclear genome and synthesized on cytosolic ribosomes, making protein import crucial for the biogenesis and function of the organelle.

Depending on their final localization, mitochondrial precursor proteins must cross either the outer or both mitochondrial membranes via dedicated protein complexes, TOM and SAM[14] in the outer membrane, and TIM22 and TIM23 in the inner membrane.[15] Most matrix-targeted precursor proteins carry an N-terminal targeting signal known as the presequence.[16,17] These precursor proteins are synthesized on 80S ribosomes in the cytosol,[18] recognized by the receptors of the TOM complex (Tom20, Tom22 and Tom70) and translocated across the outer membrane via the import pore (Tom40).[19] The presequence then interacts with the components of the TIM23 complex (Tim21 and Tim50) and traverses the inner membrane import pore (Tim23).[20,21] Further translocation of the precursor is driven by the action of the presequence-associated import motor, the PAM complex. In yeast, Ssc1 is a central component of the PAM complex, which is a highly specialized and extensively studied Hsp70/J protein machine.

Two mechanisms that explain PAM driven protein import have been proposed: the "Brownian ratchet" and the "power stroke" models.[22] Both accept precursor binding by Ssc1 to be a central feature of the process, but differ in accounts of the mechanism that pulls the precursor into the matrix. The Brownian ratchet model proposes that a precursor slides back and forth trough the Tim23 channel, with binding of Ssc1 preventing the retrograde movement resulting in the net movement toward the matrix.[23] The power stroke model stipulates a more active role for Ssc1: the chaperone exerts the force to pull the precursor through the pore.[24] The experimental data to date is insufficient to distinguish between the models; while many of the molecular interactions necessary for this complex process are understood, exact mechanistic detail remains to be elucidated.

The protein import function of the DnaK homologue Ssc1 is a novel, eukaryotic adaptation, mediated by unique proteins of the mitochondrial inner membrane. The first of these to be identified was the regulatory co-chaperone, peripheral membrane protein Tim44.[25-27] Tim44 has a matrix-exposed segment with limited similarity to a J domain, mostly in the second helix of the canonical fold. This segment of the protein does not contain the catalytic HPD motif, but is essential for Tim44 function.[28-30] Tim44 is proposed to play a double role in the initial stages of import mediated by the PAM complex: both sensing the precursor in the Tim23 channel, and recruiting Ssc1 to the TIM23 complex.[31] Tim44 can bind Ssc1 in vitro, at sites both in the ATPase and peptide-binding domain, but does not stimulate its ATPase activity, nor interact with Ssc1 in a substrate-like manner.[32-34]

A second co-chaperone associated with the inner membrane is the essential integral membrane protein Pam18, a type III J protein.[35-37] Pam18 stimulates the ATPase activity of Ssc1 enabling tight interaction between the chaperone and the precursor, necessary for the successful import reaction. The complex formed between the ADP-bound form of Ssc1and the precursor dissociates from Tim44 and is released into the matrix.[31] The nucleotide exchange factor Mge1 then mediates subsequent dissociation of the precursor from Ssc1,[38-42] with this entire cycle repeated until the precursor molecule is imported fully into the matrix.

While Tim44 is permanently associated with Tim23, Pam18 is only transiently recruited to the TIM23 complex through interactions it makes with Tim17[43] and the peripheral membrane protein Pam16.[44,45] Pam16 has a J-like domain[13] and the two proteins form a heterodimer that in vitro, and probably in vivo, persists during the stimulation of ATPase activity of Ssc1 by Pam18.[46] Pam16 acts in antagonistic manner to Pam18, reducing the stimulatory effect of Pam18 on ATP hydrolysis catalyzed by Ssc1.[47] In mitochondria, formation of the Pam16:Pam18 dimer is dependent on Pam17,[48] an integral membrane protein with no homology to known Hsp70 co-chaperones.

The J domain of Pam18 consists of only three helices, missing the additional 'fourth' helix usually found in J domains. The J-like domain of Pam16 has both sequence and predicted structural homology to helices one through three of the canonical fold, but does not contain the HPD motif. It does not stimulate the ATPase activity of Ssc1, and the addition of the HPD to the J-like domain at the end of helix II does not result in a functional J domain.[47] Recent findings indicate that Pam18 and Pam16 interact via their J and J-like domains, and this interaction determines the in vivo association and function of both proteins as part of the PAM complex.[46] This type of interactions between co-chaperones has not been observed in

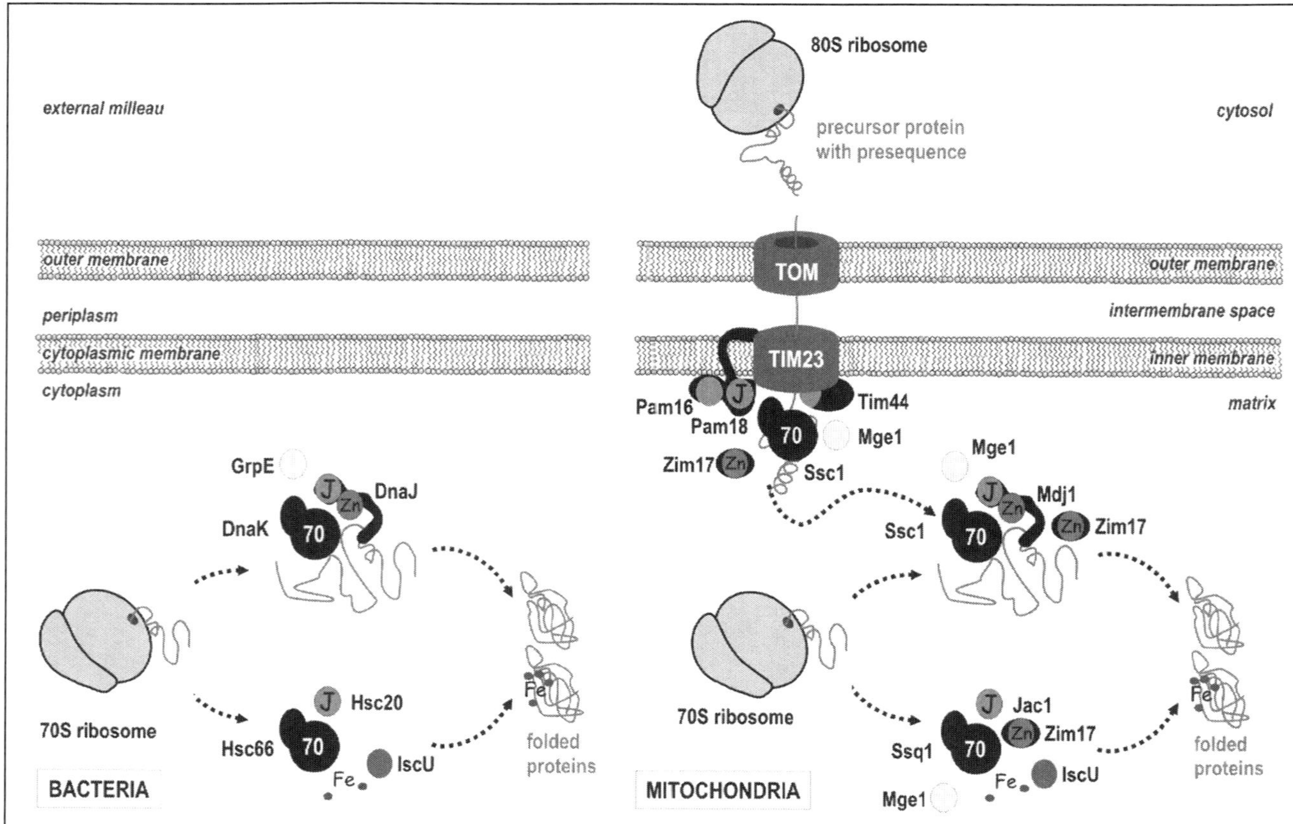

Figure 1. Co-chaperones in *Escherichia coli* and mitochondria. The figure legend is continued on the next page.

Figure 1 legend, continued. The two bacterial Hsp70 chaperone systems, mediating protein folding and FeS cluster assembly, are shown in the panel at left. The J domain (orange) and zinc finger domain (blue) of DnaJ are shown. Being a type III J protein, Hsc20 has only a J domain. DnaK also requires the action of GrpE (yellow) to efficiently complete its ATP hydrolysis cycles. In mitochondria (right) two Hsp70 isoforms, each derived from a DnaK-like ancestor, mediate protein folding and FeS cluster assembly. Each requires the assistance of the Mge1 co-chaperone. Ssq1 acts exclusively in FeS cluster assembly, while Ssc1 mediates both the translocation of proteins into the matrix and the folding of protein substrates (delivered either from the mitochondrial 70S ribosomes or from the protein import machinery). Ssc1 can function as the motor for protein import only with the assistance of the co-chaperones of the PAM complex (Pam18, Pam16), the TIM23 complex (Tim44) and Mge1. The mitochondrial co-chaperone Zim17 is required to assist Ssq1 in FeS cluster assembly, Ssc1 in protein folding and likely also the population of Ssc1 functioning in protein import: in all three processes, multiple rounds of ATP hydrolysis are required to handle a single molecule of substrate polypeptide. A color version of this figure is available online at www.eurekah.com.

any other Hsp70 system. The PAM components Sscl, Mgel, Tim44, Pam16 and Pam18 appear to be found in all eukaryotic organisms and highly conserved. In yeast, a duplicate gene (*MDJ2*) encodes a paralogue of Pam18.[49,50] It is not clear if there is any functional advantage gained from this isoform of Pam18, arisen through a relatively recent genome duplication event in yeast. However, there are several informative differences in the way the two isoforms of Pam18 perform in vitro.[50] Although the Mdj2 isoform interacts with Pam16, the latter does not antagonize, but rather enhances the Mdj2-mediated stimulation of ATP hydrolysis by Sscl. Curiously, while Pam18 contains a significant intermembrane space domain proposed to interact with the Tim17 component of the TIM23 complex, this extension is absent from Mdj2. As this domain is not essential for Pam18 function, and is absent from the Mdj2 isoform and from most Pam18 orthologs encoded on the genome sequences of other eukaryotes, its importance in PAM/TIM23 interaction seems less clear.

The number of components involved, and the unique nature of their interactions, makes the PAM complex the most complicated Hsp70 machine known to date. Its further study is likely to lead to greater understanding of the molecular and biochemical properties of Hsp70 systems in general.

Molecular Chaperones and FeS Cluster Assembly

Iron-sulfur (FeS) cluster proteins are essential cellular components found in virtually all organisms studied so far.[51] In mitochondria, they are involved in redox chemistry as components of the respiratory chain (NADH dehydrogenase, succinate dehydrogenase, Rieske protein) and metabolic conversions (aconitase, a key enzyme in the citric acid cycle). Machinery devoted to the assembly of FeS clusters is highly conserved from prokaryotes to humans;[52] in bacteria and mitochondria alike, a dedicated Hsp70/J protein system mediates this essential process.[53]

In *Escherichia coli*, the *Isc* operon encodes proteins essential for FeS cluster assembly, such as IscU[54] as well as the Hsp70 protein Hsc66[55] and the type III J protein Hsc20.[56] These two proteins interact with IscU and are important in FeS assembly in bacteria (Fig. 1), but their disruption leads to only moderate growth defects, possibly due to the existence of an alternative assembly pathway.[57-59] Hsc66 and Hsc20 have mitochondrial orthologs: in yeast these are the Hsp70 Ssq1[60] and the type III J protein Jac1.[61]

Loss of function of either Ssq1 or Jac1 results in similarly severe phenotypic consequences. Biological activity and the steady state levels of FeS proteins such as aconitase drop dramatically, accompanied by iron uptake and accumulation in mitochondria. Additionally, maturation of yeast frataxin and ferredoxin are impaired and the function of the respiratory chain function is heavily compromised.[61-65] Some of these effects are secondary consequences, rather than direct effects, with iron uptake regulated at the transcriptional level, it may be dependent on FeS assembly, rather than cellular levels of iron.

The precise role of Ssq1 and Jac1 in the assembly of FeS clusters is not yet clear. Based on iron exchange and labeling studies, these chaperones were proposed to act after the assembly of FeS clusters was complete,[66] to assist only their incorporation into holo-proteins, but their involvement in the assembly of the clusters themselves can not be ruled out. Part of the confusion stems from the fact that the only well-studied substrate of the chaperone system is the IscU homolog, the scaffold protein that accepts the nascent FeS cluster before its transfer into other apo-proteins.[58] Whether IscU homologues should be considered a substrate or an assembly factor is not entirely clear. Jac1 binds to the mitochondrial IscU protein and targets it to Ssq1 in vitro, coupling its delivery with ATPase activity stimulation. Interaction of the IscU with both Ssc1 and Jac1 is dependent on conserved residues in IscU,[67] and the IscU is fully folded when delivered to Ssq1. Ssq1 seems to mediate disassembly of the IscU/FeS cluster complex, enabling loading of the FeS cluster into the newly imported apo-enzymes. This might be just one of the many functions Ssq1 and Jac1 perform in this complex process. They have also been proposed to enforce the structure of other components of the assembly machinery, as well as preventing the aggregation of the apo-enzymes themselves. Reconstitution of steps of the assembly pathway with purified components is needed to clarify the precise role(s) Ssq1 and Jac1 play.

Given the conserved nature of the process of FeS cluster assembly between bacteria and mitochondria, the evolutionary origin of Ssq1 is rather surprising. Phylogenetic analysis reveals that it is more closely related to Ssc1 than it is to Hsc66.[68] This and other bioinformatics argues that Ssq1 and Hsc66 do not originate from the same ancestral gene, but that Ssc1 and Ssq1 are related, and share a common ancestor with DnaK. Consistent with this evolutionary scenario, biochemical differences between Hsc66 and Ssq1 are significant: Hsc66 only weakly interacts with nucleotide and does not need a nucleotide exchange factor, but Ssq1, like Ssc1, interacts strongly with nucleotides and requires Mge1 as a nucleotide exchange factor.[69] This sharing of a nucleotide exchange factor between two Hsp70 in the same cellular compartment is unique to mitochondria.[70] Disruption of Ssq1 is less detrimental to cells than depletion of Jac1, and it can be partially complemented by overexpression of Ssc1.[71] These observations suggest that Jac1, which is more abundant than Ssq1, might interact with both Ssc1 and Ssq1. If Ssc1 can also participate then Hsp70 function might be essential for FeS cluster assembly, despite dispensable nature of the Ssq1 isoform.

Zim17, a Uniquely Mitochondrial Regulator of Hsp70

Hsp70 chaperones require various co-chaperone regulators to carry out their cellular role. Since the discovery of the bacterial Hsp70 system, until recently, the only known regulators of mitochondrial Hsp70 were members of the J protein family or the nucleotide exchange factors. The first Hsp70 co-chaperone identified that does not belong to either group was the Hsp70/Hsp90 organizing protein Hop (see the Chapter in this volume by Daniels et al). Recently, a novel protein essential for function of Hsp70s has been described in mitochondria.[72] The 17 kDa peripheral membrane protein, Zim17, has an essential zinc finger domain homologous to that of bacterial DnaJ. Loss of Zim17 function leads to disruption of mitochondrial protein import, loss of function of FeS cluster proteins, aggregation of Ssc1 and Ssq1 and aberrant mitochondrial morphology.[73] Since Pam18 is a type III J protein, Zim17 was hypothesized to be the substrate-binding domain of a "fractured" co-chaperone, with the coordinated action of the two proteins resembling that of a type I J protein.[72]

Initially, it was suggest Zim17 might play a role in protein import by binding precursors emerging from the TIM23 pore and mediating their interaction with Ssc1.[72,74] Subsequent work has established that aggregation of Hsp70s (and to some extent Pam16) is the first effect of Zim17 depletion, and that all other disruptions of mitochondrial processes result from the loss of Hsp70 function.[73,75]

During its ATP hydrolysis cycle, Hsp70 can either be nucleotide-free or have either ATP or ADP bound to the ATPase domain. Nucleotide-bound forms of Hsp70 are not prone to aggregation, while nucleotide-free Hsp70 appears to aggregate. Zim17 forms a complex with nucleotide-free Ssc1 and prevents formation of nonfunctional Ssc1 oligomers.[75] There is some evidence that Zim17 interacts with the substrate binding domain of Ssc1, but the interacting surface on Zim17 has not been identified. At least to some extent, Mdj1 might be able to minimize aggregation of Ssc1 (but not Ssq1) and compensate for the loss of Zim17, since overexpression of Mdj1 minimizes the otherwise lethal consequences of downregulating the gene encoding Zim17.[72]

Homologues of Zim17 are present in all eukaryotic organisms, and all carry a mitochondrial targeting signal. No homologues have yet been identified in bacteria, or other compartments of eukaryotic cells, implying that stabilization of nucleotide-free form of Hsp70, like nucleotide exchange, might not be essential for all Hsp70 chaperones.

Concluding Remarks

Mitochondria belie their bacterial ancestry in their Hsp70 (DnaK/DnaJ/GrpE) and Hsp60 (GroEL/GroES) complement of chaperone systems. These chaperones continue to mediate the protein folding pathways that were already established in bacteria at a time before the α-proteobacterial endosymbiont ancestor of mitochondria was taken up by the first eukaryote.

The need to drive protein import, for substrate polypeptides now made externally in the cytosol, placed a demand on mitochondria that has been met with a series of novel co-chaperones. The uniquely eukaryotic proteins Pam16, Pam18 and Tim44 enabled the existing Hsp70 (Ssc1) to be recruited as a protein import motor. That this motor is truly ubiquitous in eukaryotes is made certain from the recent finding of Pam18 in the anaerobic protists *Giardia intestinalis* and *Trichomonas vaginalis*.[76]

In the course of evolution, the progenitor DnaK-type chaperone has been modified into two forms, one isoform (Ssc1) mediating protein import and protein folding and a second isoform (Ssq1) that mediates FeS cluster assembly. It is intriguing that a uniquely mitochondrial co-chaperone, Zim17, has been created to stabilize both Ssc1 and Ssq1. It is not clear why the mitochondrial Hsp70s require their structure to be enforced in a manner so distinct from other Hsp70s.

References

1. Rowley N, Prip-Buus C, Westermann B et al. Mdj1p, a novel chaperone of the DnaJ family, is involved in mitochondrial biogenesis and protein folding. Cell 1994; 77(2):249-259.
2. Bolliger L, Deloche O, Glick BS et al. A mitochondrial homolog of bacterial GrpE interacts with mitochondrial hsp70 and is essential for viability. EMBO J 1994; 13(8):1998-2006.
3. Prip-Buus C, Westerman B, Schmitt M et al. Role of the mitochondrial DnaJ homologue, Mdj1p, in the prevention of heat-induced protein aggregation. FEBS Lett 1996; 380(1-2):142-146.
4. Kubo Y, Tsunehiro T, Nishikawa S et al. Two distinct mechanisms operate in the reactivation of heat-denatured proteins by the mitochondrial Hsp70/Mdj1p/Yge1p chaperone system. J Mol Biol 1999; 286(2):447-464.
5. Deloche O, Liberek K, Zylicz M et al. Purification and biochemical properties of Saccharomyces cerevisiae Mdj1p, the mitochondrial DnaJ homologue. J Biol Chem 1997; 272(45):28539-28544.
6. Deloche O, Kelley WL, Georgopoulos C. Structurefunction analyses of the Ssc1p, Mdj1p, and Mge1p Saccharomyces cerevisiae mitochondrial proteins in Escherichia coli. J Bacteriol 1997; 179(19):6066-6075.
7. Liu Q, Krzewska J, Liberek K et al. Mitochondrial Hsp70 Ssc1: Role in protein folding. J Biol Chem 2001; 276(9):6112-6118.
8. Horst M, Oppliger W, Rospert S et al. Sequential action of two hsp70 complexes during protein import into mitochondria. EMBO J 1997; 16(8):1842-1849.
9. Heyrovska N, Frydman J, Hohfeld J et al. Directionality of polypeptide transfer in the mitochondrial pathway of chaperone-mediated protein folding. Biol Chem 1998; 379(3):301-309.
10. Manning-Krieg UC, Scherer PE, Schatz G. Sequential action of mitochondrial chaperones in protein import into the matrix. EMBO J 1991; 10(11):3273-3280.
11. Westermann B, Gaume B, Herrmann JM et al. Role of the mitochondrial DnaJ homolog Mdj1p as a chaperone for mitochondrially synthesized and imported proteins. Mol Cell Biol 1996; 16(12):7063-7071.

12. Duchniewicz M, Germaniuk A, Westermann B et al. Dual role of the mitochondrial chaperone Mdj1p in inheritance of mitochondrial DNA in yeast. Mol Cell Biol 1999; 19(12):8201-8210.
13. Walsh P, Bursac D, Law YC et al. The J-protein family: Modulating protein assembly, disassembly and translocation. EMBO Rep 2004; 5(6):567-571.
14. Pfanner N, Wiedemann N, Meisinger C et al. Assembling the mitochondrial outer membrane. Nat Struct Mol Biol 2004; 11(11):1044-1048.
15. Truscott KN, Brandner K, Pfanner N. Mechanisms of protein import into mitochondria. Curr Biol 2003; 13(8):R326-337.
16. Pfanner N, Geissler A. Versatility of the mitochondrial protein import machinery. Nat Rev Mol Cell Biol 2001; 2(5):339-349.
17. Pfanner N. Protein sorting: Recognizing mitochondrial presequences. Curr Biol 2000; 10(11):R412-415.
18. Lithgow T. Targeting of proteins to mitochondria. FEBS Lett 2000; 476(1-2):22-26.
19. Hill K, Model K, Ryan MT et al. Tom40 forms the hydrophilic channel of the mitochondrial import pore for preproteins [see comment]. Nature 1998; 395(6701):516-521.
20. Koehler CM. New developments in mitochondrial assembly. Annu Rev Cell Dev Biol 2004; 20:309-335.
21. Jensen RE, Dunn CD. Protein import into and across the mitochondrial inner membrane: Role of the TIM23 and TIM22 translocons. Biochim Biophys Acta 2002; 1592(1):25-34.
22. Mayer MP. Timing the catch. Nat Struct Mol Biol 2004; 11(1):6-8.
23. Liu Q, D'Silva P, Walter W et al. Regulated cycling of mitochondrial Hsp70 at the protein import channel. Science 2003; 300(5616):139-141.
24. Voos W, von Ahsen O, Muller H et al. Differential requirement for the mitochondrial Hsp70-Tim44 complex in unfolding and translocation of preproteins. EMBO J 1996; 15(11):2668-2677.
25. Maarse AC, Blom J, Grivell LA et al. MPI1, an essential gene encoding a mitochondrial membrane protein, is possibly involved in protein import into yeast mitochondria. EMBO J 1992; 11(10):3619-3628.
26. Scherer PE, Manning-Krieg UC, Jeno P et al. Identification of a 45-kDa protein at the protein import site of the yeast mitochondrial inner membrane. Proc Natl Acad Sci USA 1992; 89(24):11930-11934.
27. Weiss C, Oppliger W, Vergeres G et al. Domain structure and lipid interaction of recombinant yeast Tim44. Proc Natl Acad Sci USA 1999; 96(16):8890-8894.
28. Merlin A, Voos W, Maarse AC et al. The J-related segment of tim44 is essential for cell viability: A mutant Tim44 remains in the mitochondrial import site, but inefficiently recruits mtHsp70 and impairs protein translocation. J Cell Biol 1999; 145(5):961-972.
29. Rassow J, Dekker PJ, van Wilpe S et al. The preprotein translocase of the mitochondrial inner membrane: Function and evolution. J Mol Biol 1999; 286(1):105-120.
30. Rassow J, Maarse AC, Krainer E et al. Mitochondrial protein import: Biochemical and genetic evidence for interaction of matrix hsp70 and the inner membrane protein MIM44. J Cell Biol 1994; 127(6 Pt 1):1547-1556.
31. D'Silva P, Liu Q, Walter W et al. Regulated interactions of mtHsp70 with Tim44 at the translocon in the mitochondrial inner membrane. Nat Struct Mol Biol 2004; 11(11):1084-1091.
32. Krimmer T, Rassow J, Kunau WH et al. Mitochondrial protein import motor: The ATPase domain of matrix Hsp70 is crucial for binding to Tim44, while the peptide binding domain and the carboxy-terminal segment play a stimulatory role. Mol Cell Biol 2000; 20(16):5879-5887.
33. Moro F, Okamoto K, Donzeau M et al. Mitochondrial protein import: Molecular basis of the ATP-dependent interaction of MtHsp70 with Tim44. J Biol Chem 2002; 277(9):6874-6880.
34. Strub A, Rottgers K, Voos W. The Hsp70 peptide-binding domain determines the interaction of the ATPase domain with Tim44 in mitochondria. EMBO J 2002; 21(11):2626-2635.
35. D'Silva PD, Schilke B, Walter W et al. J protein co-chaperone of the mitochondrial inner membrane required for protein import into the mitochondrial matrix. Proc Natl Acad Sci USA 2003; 100(24):13839-13844.
36. Truscott KN, Voos W, Frazier AE et al. A J-protein is an essential subunit of the presequence translocase-associated protein import motor of mitochondria. J Cell Biol 2003; 163(4):707-713.
37. Mokranjac D, Sichting M, Neupert W et al. Tim14, a novel key component of the import motor of the TIM23 protein translocase of mitochondria. EMBO J 2003; 22(19):4945-4956.
38. Laloraya S, Dekker PJ, Voos W et al. Mitochondrial GrpE modulates the function of matrix Hsp70 in translocation and maturation of preproteins. Mol Cell Biol 1995; 15(12):7098-7105.
39. Laloraya S, Gambill BD, Craig EA. A role for a eukaryotic GrpE-related protein, Mge1p, in protein translocation. Proc Natl Acad Sci USA 1994; 91(14):6481-6485.

40. Westermann B, Prip-Buus C, Neupert W et al. The role of the GrpE homologue, Mge1p, in mediating protein import and protein folding in mitochondria. EMBO J 1995; 14(14):3452-3460.
41. Voos W, Gambill BD, Laloraya S et al. Mitochondrial GrpE is present in a complex with hsp70 and preproteins in transit across membranes. Mol Cell Biol 1994; 14(10):6627-6634.
42. Schneider HC, Westermann B, Neupert W et al. The nucleotide exchange factor MGE exerts a key function in the ATP-dependent cycle of mt-Hsp70-Tim44 interaction driving mitochondrial protein import. EMBO J 1996; 15(21):5796-5803.
43. Chacinska A, Lind M, Frazier AE et al. Mitochondrial presequence translocase: Switching between TOM tethering and motor recruitment involves Tim21 and Tim17. Cell 2005; 120(6):817-829.
44. Frazier AE, Dudek J, Guiard B et al. Pam16 has an essential role in the mitochondrial protein import motor. Nat Struct Mol Biol 2004; 11(3):226-233.
45. Kozany C, Mokranjac D, Sichting M et al. The J domain-related co-chaperone Tim16 is a constituent of the mitochondrial TIM23 preprotein translocase. Nat Struct Mol Biol 2004; 11(3):234-241.
46. D'Silva PR, Schilke B, Walter W et al. Role of Pam16's degenerate J domain in protein import across the mitochondrial inner membrane. Proc Natl Acad Sci USA 2005; 102(35):12419-12424.
47. Li Y, Dudek J, Guiard B et al. The presequence translocase-associated protein import motor of mitochondria. Pam16 functions in an antagonistic manner to Pam18. J Biol Chem 2004; 279(36):38047-38054.
48. van der Laan M, Chacinska A, Lind M et al. Pam17 is required for architecture and translocation activity of the mitochondrial protein import motor. Mol Cell Biol 2005; 25(17):7449-7458.
49. Westermann B, Neupert W. Mdj2p, a novel DnaJ homolog in the mitochondrial inner membrane of the yeast Saccharomyces cerevisiae. J Mol Biol 1997; 272(4):477-483.
50. Mokranjac D, Sichting M, Popov-Celeketic D et al. The import motor of the yeast mitochondrial TIM23 preprotein translocase contains two different J proteins, Tim14 and Mdj2. J Biol Chem 2005; 280(36):31608-31614.
51. Balk J, Lill R. The cell's cookbook for iron-sulfur clusters: Recipes for fool's gold? Chembiochem 2004; 5(8):1044-1049.
52. Muhlenhoff U, Lill R. Biogenesis of iron-sulfur proteins in eukaryotes: A novel task of mitochondria that is inherited from bacteria. Biochim Biophys Acta 2000; 1459(2-3):370-382.
53. Craig EA, Marszalek J. A specialized mitochondrial molecular chaperone system: A role in formation of Fe/S centers. Cell Mol Life Sci 2002; 59(10):1658-1665.
54. Agar JN, Krebs C, Frazzon J et al. IscU as a scaffold for iron-sulfur cluster biosynthesis: Sequential assembly of [2Fe-2S] and [4Fe-4S] clusters in IscU. Biochemistry 2000; 39(27):7856-7862.
55. Seaton BL, Vickery LE. A gene encoding a DnaK/hsp70 homolog in Escherichia coli. Proc Natl Acad Sci USA 1994; 91(6):2066-2070.
56. Kawula TH, Lelivelt MJ. Mutations in a gene encoding a new Hsp70 suppress rapid DNA inversion and bgl activation, but not proU derepression, in hns-1 mutant Escherichia coli. J Bacteriol 1994; 176(3):610-619.
57. Takahashi Y, Tokumoto U. A third bacterial system for the assembly of iron-sulfur clusters with homologs in archaea and plastids. J Biol Chem 2002; 277(32):28380-28383.
58. Frazzon J, Dean DR. Formation of iron-sulfur clusters in bacteria: An emerging field in bioinorganic chemistry. Curr Opin Chem Biol 2003; 7(2):166-173.
59. Tokumoto U, Takahashi Y. Genetic analysis of the isc operon in Escherichia coli involved in the biogenesis of cellular iron-sulfur proteins. J Biochem (Tokyo) 2001; 130(1):63-71.
60. Knight SA, Sepuri NB, Pain D et al. Mt-Hsp70 homolog, Ssc2p, required for maturation of yeast frataxin and mitochondrial iron homeostasis. J Biol Chem 1998; 273(29):18389-18393.
61. Voisine C, Cheng YC, Ohlson M et al. Jac1, a mitochondrial J-type chaperone, is involved in the biogenesis of Fe/S clusters in Saccharomyces cerevisiae. Proc Natl Acad Sci USA 2001; 98(4):1483-1488.
62. Lutz T, Westermann B, Neupert W et al. The mitochondrial proteins Ssq1 and Jac1 are required for the assembly of iron sulfur clusters in mitochondria. J Mol Biol 2001; 307(3):815-825.
63. Schilke B, Voisine C, Beinert H et al. Evidence for a conserved system for iron metabolism in the mitochondria of Saccharomyces cerevisiae. Proc Natl Acad Sci USA 1999; 96(18):10206-10211.
64. Voisine C, Schilke B, Ohlson M et al. Role of the mitochondrial Hsp70s, Ssc1 and Ssq1, in the maturation of Yfh1. Mol Cell Biol 2000; 20(10):3677-3684.
65. Kim R, Saxena S, Gordon DM et al. J-domain protein, Jac1p, of yeast mitochondria required for iron homeostasis and activity of Fe-S cluster proteins. J Biol Chem 2001; 276(20):17524-17532.
66. Muhlenhoff U, Gerber J, Richhardt N et al. Components involved in assembly and dislocation of iron-sulfur clusters on the scaffold protein Isu1p. EMBO J 2003; 22(18):4815-4825.

67. Dutkiewicz R, Schilke B, Cheng S et al. Sequence-specific interaction between mitochondrial Fe-S scaffold protein Isu and Hsp70 Ssq1 is essential for their in vivo function. J Biol Chem 2004; 279(28):29167-29174.
68. Huynen MA, Snel B, Bork P et al. The phylogenetic distribution of frataxin indicates a role in iron-sulfur cluster protein assembly. Hum Mol Genet 2001; 10(21):2463-2468.
69. Dutkiewicz R, Schilke B, Knieszner H et al. Ssq1, a mitochondrial Hsp70 involved in iron-sulfur (Fe/S) center biogenesis. Similarities to and differences from its bacterial counterpart. J Biol Chem 2003; 278(32):29719-29727.
70. Schmidt S, Strub A, Rottgers K et al. The two mitochondrial heat shock proteins 70, Ssc1 and Ssq1, compete for the co-chaperone Mge1. J Mol Biol 2001; 313(1):13-26.
71. Schilke B, Forster J, Davis J et al. The cold sensitivity of a mutant of Saccharomyces cerevisiae lacking a mitochondrial heat shock protein 70 is suppressed by loss of mitochondrial DNA. J Cell Biol 1996; 134(3):603-613.
72. Burri L, Vascotto K, Fredersdorf S et al. Zim17, a novel zinc finger protein essential for protein import into mitochondria. J Biol Chem 2004; 279(48):50243-50249.
73. Sanjuan Szklarz LK, Guiard B, Rissler M et al. Inactivation of the mitochondrial heat shock protein zim17 leads to aggregation of matrix hsp70s followed by pleiotropic effects on morphology and protein biogenesis. J Mol Biol 2005; 351(1):206-218.
74. Yamamoto H, Momose T, Yatsukawa Y et al. Identification of a novel member of yeast mitochondrial Hsp70-associated motor and chaperone proteins that facilitates protein translocation across the inner membrane. FEBS Lett 2005; 579(2):507-511.
75. Sichting M, Mokranjac D, Azem A et al. Maintenance of structure and function of mitochondrial Hsp70 chaperones requires the chaperone Hep1. EMBO J 2005; 24(5):1046-1056.
76. Dolezal P, Smid O, Rada P et al. Giardia mitosomes and trichomonad hydrogenosomes share a common mode of protein targeting. Proc Natl Acad Sci USA 2005; 102(31):10924-10929.

Chapter 10

From Creator to Terminator:

Co-Chaperones That Link Molecular Chaperones to the Ubiquitin/Proteasome System

Jörg Höhfeld,* Karsten Böhse, Markus Genau and Britta Westhoff

Abstract

Molecular chaperones are well known as intracellular mediators of protein folding. An active participation in protein degradation only recently emerged from the functional characterization of certain co-chaperones. In the light of these novel findings long held views regarding the interplay of chaperones and proteases in protein quality control need to be reconsidered. A further elucidation of chaperone-assisted degradation will be essential to understand the molecular basis of protein homeostasis.

Abbreviations

BAG: BAG-1 homology domain; BAG-1: Bcl2-associated athanogene 1; BAG-2: Bcl2-associated athanogene 2; BAG-5: Bcl2-associated athanogene 5; BTB: broad complex, tramtrack and bric a brac domain; CFTR: cystic fibrosis transmembrane conductance regulator; CHIP: carboxyl terminus of Hsp70-interacting protein; E1: ubiquitin-activating enzyme; E2: ubiquitin-conjugating enzyme; E3: ubiquitin ligase; HECT: homologous to E6AP carboxyl terminus; Hip: Hsp70-interacting protein; Hop: Hsp70/Hsp90-organizing protein; HspBP1: Hsp70-binding protein 1; HSJ1: Homo sapiens J protein 1; RING: really interesting new gene; TPR: tetratricopeptide repeat; U-box: UFD2 homology domain; UBA: ubiquitin associated domain; UBC: ubiquitin-conjugating enzyme; UBL: ubiquitin-like domain; UIM: ubiquitin-interaction motif.

Introduction

It is textbook knowledge that molecular chaperones mediate intracellular protein folding. Their ability to bind and stabilize nonnative conformations of newly synthesized or damaged proteins enables molecular chaperones to facilitate the adoption of the native, biologically active structure.[1-3] The same ability, however, makes molecular chaperones ideally suited to assist protein degradation. By maintaining misfolded or aggregation-prone proteins in a soluble state chaperones could ensure recognition by cellular degradation machineries, such as the ubiquitin/proteasome system, and promote proper disposal. In recent years more and more data have emerged that support a degradation function of at least some molecular chaperones, e.g., members of the Hsp70 and Hsp90 chaperone families.[4-6] Their cooperation with the

*Corresponding Author: Jörg Höhfeld—Institut für Zellbiologie und Bonner Forum Biomedizin, Rheinische Friedrich-Wilhelms Universität Bonn, Ulrich-Haberland-Str. 61a, D-53121 Bonn, Germany. Email: hoehfeld@uni-bonn.de

Networking of Chaperones by Co-Chaperones, edited by Gregory L. Blatch.

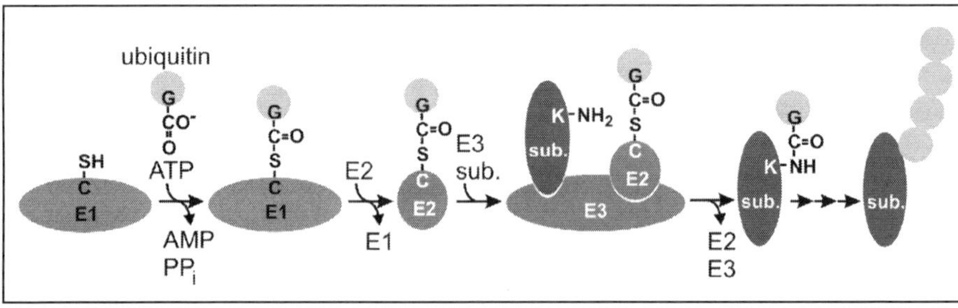

Figure 1. Mechanism of ubiquitin conjugation. ATP-dependent activation of ubiquitin is mediated by the E1 enzyme and involves the formation of a thioester bond between the C-terminal glycine of ubiquitin and a cysteine residue of E1. The activated ubiquitin is transferred onto a cysteine residue of the ubiquitin-conjugating E2 enzyme, which finally cooperates with an E3 ubiquitin ligase in the attachment of ubiquitin to a lysine residue of the protein substrate via an isopeptide bond. In the case of HECT-type E3s ubiquitylation involves the formation of a thioester bond between the ligase and activated ubiquitin (not shown). Lysine residues of the conjugated ubiquitin are used for chain formation in subsequent reactions. A color version of this figure is available online at www.eurekah.com.

ubiquitin/proteasome system is mediated by a set of dedicated co-chaperones. In the following we will describe these co-chaperones, the functional characterization of which has significantly expanded our understanding of intracellular protein quality control. At the same time novel questions arise. What regulates the balance between chaperone-assisted protein folding and degradation? Is the chaperone system able to discriminate between nonnative proteins doomed for degradation and those that need to be refolded? Answers to these questions have only begun to emerge. Addressing them may pave the way to therapeutic modulation of chaperone pathways in neurodegenerative diseases and cancer in the future.

The Ubiquitin/Proteasome System

Before we will describe how chaperone activity can be switched from protein folding to protein degradation—how the creator is turned into a terminator—it appears necessary to introduce the ubiquitin/proteasome system, which is the main degradation machinery for the removal of misfolded and short lived proteins in the eukaryotic cytoplasm and nucleus and which mediates ER-associated degradation.[7,8] As the name says the system comprises two main components: (i) ubiquitin—a small protein of 76 amino acids that is expressed in all eukaryotic cells and serves as a degradation signal when conjugated onto other proteins in the form of a polyubiquitin chain,[9] and (ii) the proteasome—a large oligomeric protein complex with a central proteolytic cavity in which polyubiquitylated proteins can be degraded in a manner separated from the cellular surrounding.[10] This brief description already points to distinct steps during the degradation process. The protein doomed for destruction has to be modified by ubiquitin chain attachment, which requires the activation of ubiquitin and the specific recognition of the protein substrate by a conjugation machinery, followed by sorting to the proteasome and proteolytic cleavage inside the proteasome cavity. Ubiquitin activation is mediated by a single ubiquitin-activating enzyme, termed E1, and involves the formation of a thioester bond between the C-terminal glycine of ubiquitin and a cysteine residue of the enzyme (Fig. 1). The activated ubiquitin is transferred onto the E2 ubiquitin-conjugating enzyme involving again thioester bond formation, before covalent attachment to lysine residues of the protein substrate is assisted by an E3 ubiquitin ligase.[9,11] Lysine residues of the attached ubiquitin itself are subsequently used for the conjugation of additional ubiquitin moieties, leading to chain formation (Fig. 1). A lysine-48 linked chain usually serves as the degradation signal. In some instances, chain formation requires additional proteins that cooperate with the E2/E3 machinery.[12] Thirty-four distinct E2s for ubiquitin conjugation are present in the human genome, all

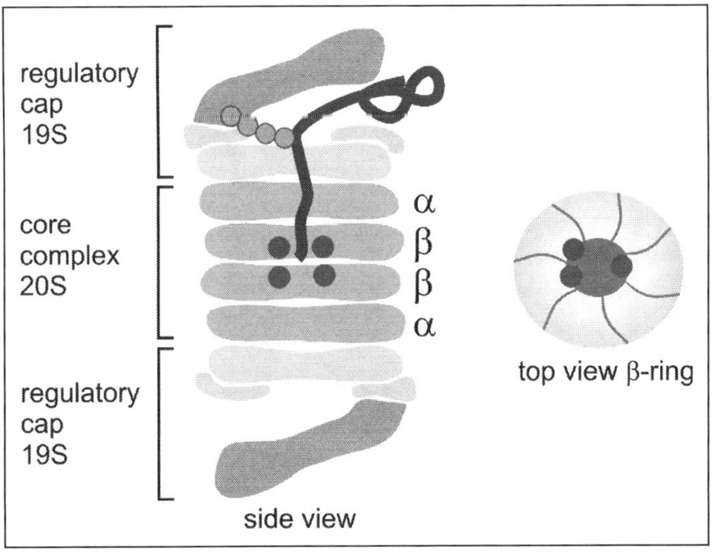

Figure 2. Schematic presentation of the 26S proteasome, which consists of 19S regulatory caps attached to both ends of the 20S catalytic core particle. The latter is formed by four heptameric rings of α and β subunits stacked onto each other in an αββα arrangement. Only some β subunits are proteolytically active and their catalytic residues face the inner lining of the core particle (red dots). Subunits of the regulatory cap mediate recognition of ubiquitin chains, substrate unfolding and deubiquitylation to allow for the insertion of the protein substrate into the proteolytic core. A color version of this figure is available online at www.eurekah.com.

defined by the presence of a signature UBC domain (H. Scheel and K. Hofmann, Memorec, Cologne, personal communication). E3s are recruited from diverse protein families containing for example RING, HECT, U-box or BTB domains. The fact that there are 268 different RING proteins encoded in the human genome (H. Scheel and K. Hofmann, personal communication) already provides a glimpse at the high degree of diversity that is achieved at the level of E2/E3 pairing. This diversity enables the conjugation machinery to recognize thousands of different protein substrates in a specific manner.

The attachment of a lysine-48 linked ubiquitin chain seems to provide sufficient targeting information to direct a protein substrate to the proteasome.[13] The ubiquitin chain is recognized by receptor proteins present in the 19S regulatory cap of the proteasome and insertion of the substrate into the 20S core particle is initiated (Fig. 2).[10] Nevertheless, several proteins were recently shown to assist proteasomal sorting.[14-16] These proteins often possess ubiquitin-binding (UBA, UIM) domains that enable them to associate with ubiquitylated proteins and to accompany them during proteasomal sorting in cooperation with dedicated chaperones. Often the escort proteins also expose ubiquitin-like (UBL) domains, which are recognized by specialized subunits of the regulatory cap to facilitate docking at the proteasome.[15] Besides providing additional means to regulate proteasomal degradation, the employment of escort proteins seems to reflect a requirement for chaperoning ubiquitylated proteins during the sorting process. A cooperation of molecular chaperones with the ubiquitin/proteasome system may thus occur at two distinct stages of ubiquitin-mediated degradation. Chaperones could present misfolded and aggregation-prone proteins to specialized E2/E3 ubiquitylation machineries during initial substrate selection, but could also cooperate with escort proteins to prevent the aggregation of ubiquitylated substrates on the sorting pathway. As we will see evidence for both mechanisms is emerging from the functional characterization of certain co-chaperones.

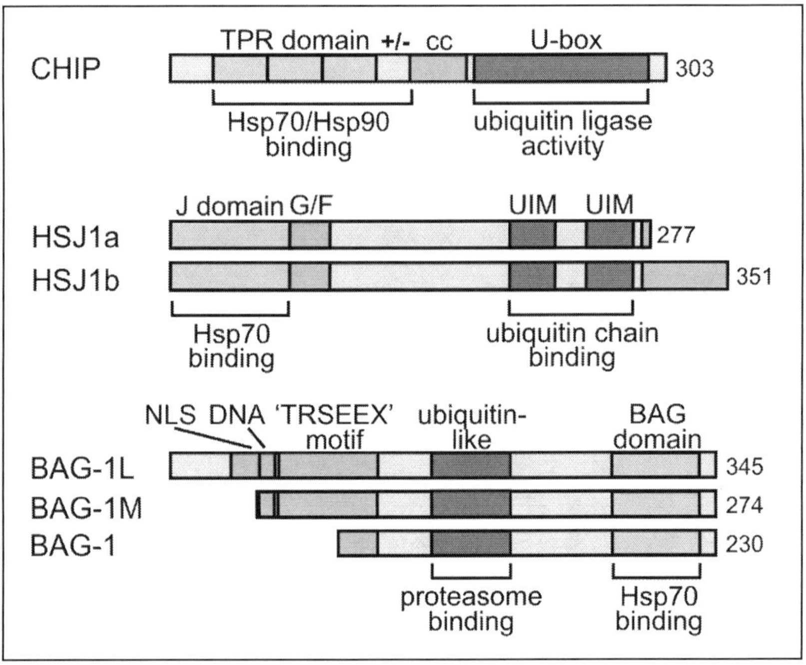

Figure 3. Co-chaperones that link molecular chaperones to the ubiquitin/proteasome system. CHIP, HSJ1 and BAG-1 combine chaperone binding sites with functional domains involved in ubiquitin-mediated degradation. TPR: tetratricopeptide repeat; +/-: highly charged region of CHIP involved in chaperone binding; cc: coiled coil domain; U-box: UFD2 homology domain; G/F: glycine and phenylalanine rich region; UIM: ubiquitin interaction motif; NLS: nuclear localization signal; DNA: DNA binding region; 'TRSEEX': region carrying multiple repeats of the pentapeptide TRSEEX; BAG: BAG-1 homology domain. A color version of this figure is available online at www.eurekah.com.

Co-Chaperones That Link Chaperones to the Ubiquitin/Proteasome System

A subset of co-chaperones apparently evolved to mediate a cooperation of the molecular chaperones Hsp70 and Hsp90 with the ubiquitin/proteasome system.[6] These co-chaperones possess chaperone-binding sites together with protein domains that act in ubiquitin-mediated degradation (Fig. 3). As our knowledge about functional protein domains increases everyday, more and more such co-chaperones emerge. The few examples studied in detail to date suggest a close and highly regulated interplay between chaperones and the degradation machinery, which were long considered as opposing forces in controlling protein biogenesis.

CHIP—A Chaperone-Associated Ubiquitin Ligase

The carboxyl terminus of Hsp70-interacting protein, CHIP, was identified in 1999 by the Patterson laboratory in a screen for novel co-chaperones expressed in the heart.[17] Expression of CHIP, however, is not restricted to the heart. It is broadly expressed in diverse organs and tissues.[17] In addition to a tetratricopeptide-repeat (TPR) domain that mediates binding to the carboxyl termini of Hsp70 and Hsp90,[17,18] CHIP possesses a U-box that confers ubiquitin ligase activity to the co-chaperone (Fig. 3).[19-21] The U-box enables the co-chaperone to cooperate with ubiquitin-conjugating enzymes of the Ubc4/5 family in the ubiquitylation

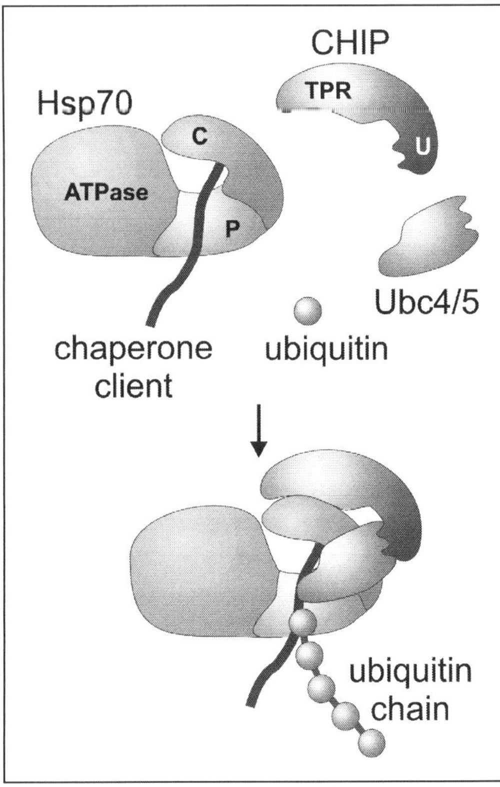

Figure 4. Formation of the Hsp70/CHIP complex, which functions as a multi subunit ubiquitin ligase. CHIP uses its TPR domain to interact with the carboxyl terminus of Hsp70 (C) that forms a lid over the peptide binding domain (P). The U-box of CHIP recruits E2 ubiquitin-conjugating enzymes of the Ubc4/5 family to mediate the ubiquitylation of the chaperone-bound client. ATPase: amino terminal ATPase domain of Hsp70. A color version of this figure is available online at www.eurekah.com.

of chaperone clients (Fig. 4). Affected clients include (i) proteins that are recognized by the chaperone systems during protein quality control, such as immature forms of the cystic fibrosis transmembrane conductance regulator (CFTR) and hyperphosphorylated forms of the tau protein associated with Alzheimer's disease,[22-27] and (ii) proteins that undergo chaperone-assisted conformational changes during their cellular function, for example the glucocorticoid hormone receptor.[18,19] Furthermore, apoptosis regulators, i.e., the tumor suppressor p53 and the apoptosis signal-regulating kinase 1, were recently added to the growing list of CHIP substrates.[28,29] Taken together, a central role of CHIP in protein quality control and the regulation of signal transduction and apoptosis becomes evident.

The ability of CHIP to mediate ubiquitylation and degradation depends on an intact TPR domain as well as a functional U-box.[18,22] This demonstrates that CHIP needs to cooperate with either Hsp70 or Hsp90 during client recognition. The chaperones most likely present a bound client protein to the CHIP conjugation machinery (Fig. 4). According to this model the chaperone/CHIP complex might be viewed as a multi subunit ubiquitin ligase, in which the chaperones act as main substrate recognition factors and CHIP switches chaperone activity from protein folding to protein degradation.

HSJ1—A Neuronal Escort Protein for the Sorting of Chaperone Clients to the Proteasome

HSJ1 is a co-chaperone that is predominantly expressed in neuronal cells.[30] Two isoforms can be distinguished that differ with regard to their cellular localization. HSJ1a is localized in the cytoplasm, whereas HSJ1b is recruited to the cytoplasmic face of the endoplasmic reticulum (ER) due to lipid attachment at its extended carboxyl terminus (Fig. 3). Both isoforms belong to the family of DnaJ/Hsp40 co-chaperones that contain a J-domain to stimulate ATP hydrolysis of Hsp70 proteins.[2,31] Similar to other J-proteins (see Chapter by Rosser and Cyr), HSJ1 possesses an intrinsic chaperone activity.[16] This enables the co-chaperone to deliver chaperone clients to Hsp70. Efficient loading of the client onto the chaperone is subsequently facilitated by the ATPase-stimulating activity of HSJ1, which drives Hsp70 in the ADP conformation with high client binding affinity (Fig. 5). Of interest with regard to chaperone-assisted

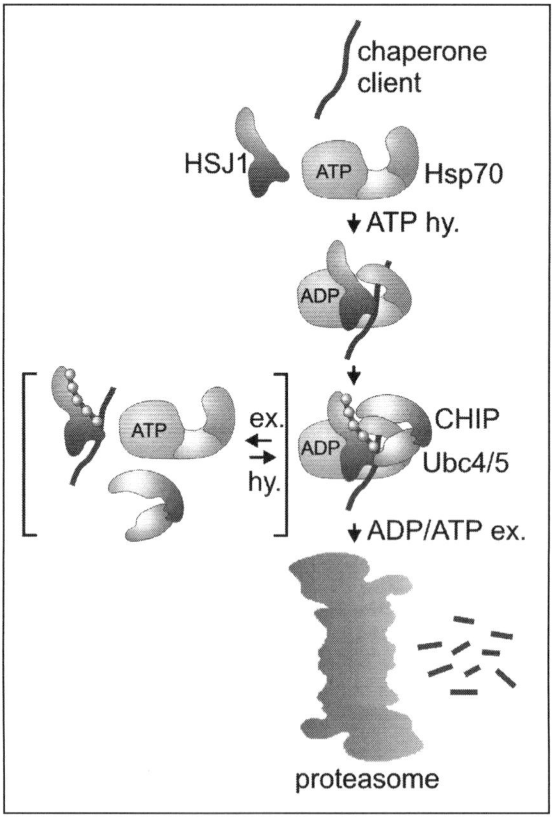

Figure 5. Model for the cooperation of HSJ1 with CHIP during chaperone-assisted degradation in neuronal cells. HSJ1 stimulates ATP hydrolysis on Hsp70 (hy.) to facilitate client loading onto the chaperone. The CHIP ubiquitin conjugation machinery associates with the formed chaperone complex and ubiquitylates the chaperone client. Once ubiquitylated the client is protected against chain trimming by ubiquitin hydrolases due to the UIM-mediated interaction of HSJ1 with the attached ubiquitin chain. Upon ADP/ATP exchange (ex.) the client is released and inserted into the proteasome for degradation. If nucleotide exchange occurs prior to docking at the proteasome, HSJ1 remains associated with the ubiquitylated client and stimulates reloading onto Hsp70 (brackets). For sake of clarity ubiquitylation of HSJ1 and Hsp70 was omitted. A color version of this figure is available online at www.eurekah.com.

degradation is the fact that HSJ1 carries two UIM domains utilized for binding ubiquitin chains (Fig. 3).[16] HSJ1 may thus be viewed as an Hsp70-loading factor for ubiquitylated chaperone clients. Indeed, a close cooperation between HSJ1 and the CHIP ubiquitin ligase was recently observed.[16] HSJ1 stimulates CHIP mediated ubiquitylation and then binds the assembled ubiquitin chain via its UIM domains. In association with HSJ1 the ubiquitin chain is protected against the activity of ubiquitin hydrolases that are usually employed to truncate ubiquitin chains as a regulatory measure during ubiquitin-mediated degradation. By inhibiting chain disassembly HSJ1 further facilitates proteasomal sorting. The Hsp70-loading activity of HSJ1 finally ensures that Hsp70 remains engaged in the sorting of the ubiquitylated chaperone client until docking at the proteasome is achieved (Fig. 5). HSJ1 apparently fulfils an escort function during the sorting of chaperone clients to the proteasome in the cytoplasm and at the ER membrane.[16] The characterization of HSJ1 is of particular relevance with regard to neurodegenerative diseases pathologically defined by the accumulation of protein aggregates, such as Parkinson's and Huntington's disease.[32] An elevation of HSJ1 levels in neuronal cells, indeed, reduces the aggregation and stimulates the degradation of a disease-causing fragment of the Huntingtin protein.[16] The neuronal expression of HSJ1 may therefore be considered as a protective mechanism to cope with cytotoxic protein aggregation in postmitotic cells. Consequently, HSJ1 may represent an interesting therapeutic target for the treatment of neurodegenerative diseases.

BAG-1—A Nucleotide Exchange Factor of Hsc70 That Binds to the Proteasome

BAG-1 (Bcl-2 associated athanogene 1) was initially identified as an interaction partner of the anti-apoptotic Bcl-2 protein and only subsequently shown to bind and regulate Hsp70.[33-35] How these findings correlate is still unclear. Three main isoforms of BAG-1 have been detected, which are expressed at various levels in diverse tissues (Fig. 3).[36] Moreover, the co-chaperone is the founding member of a BAG domain-containing protein family.[37] The BAG domain mediates binding to the ATPase domain of Hsp70 and accelerates ADP/ATP exchange.[35,38] As a consequence the release of client proteins from Hsp70 is stimulated. Intriguingly, BAG-1 also possesses a UBL domain that is utilized for an interaction with the proteasome (Fig. 6).[39] The UBL domain is recognized by the Rpn1 subunit of the 19S regulatory cap (C. Gordon, MRC, Edinburgh, personal communication). Due to its domain architecture the co-chaperone apparently provides a physical link between Hsp70 and the proteolytic complex. Accordingly, elevation of BAG-1 levels in cell culture experiments results in an increased association of the chaperone with the proteasome.[39] In summary, the data indicate that BAG-1 fulfils its cellular function at least in part by stimulating client unloading from Hsp70 at the proteasome. Indeed, BAG-1 promotes CHIP-mediated turnover of the glucocorticoid hormone receptor.[19] A cooperation of the two co-chaperones in chaperone-assisted degradation is possible because they occupy different domains on Hsp70. BAG-1 binds to the ATPase domain, whereas CHIP simultaneously associates with the carboxyl terminus of Hsp70, resulting in the formation of a chaperone complex dedicated for degradation (Fig. 6).

BAG-1, HSJ1 and also Hsp70 are themselves substrates of the CHIP ubiquitin ligase.[16,20,40] During chaperone-assisted degradation they become ubiquitylated by CHIP. However, CHIP does not trigger the proteasomal degradation of these components. Ubiquitylation of the co-chaperones and Hsp70 seems to provide an additional means to facilitate docking at the proteasome during the delivery of chaperone clients (Fig. 6). In the case of BAG-1, it was indeed shown that CHIP-mediated ubiquitylation stimulates binding to the proteasome.[40] The chaperone/co-chaperone complex apparently exposes multiple proteasomal sorting signals. This mirrors the recent identification of multiple receptor proteins for the recognition of ubiquitin-like domains and ubiquitin chains within the 19S regulatory cap of the proteasome (Fig. 6).[15,41] Using multiple docking sites might be necessary during CHIP-induced degradation because CHIP in cooperation with Ubc4/5 does not mediate the assembly of lysine-48 linked ubiquitin chains, which are most efficiently recognized at the proteasome. Instead, mixed

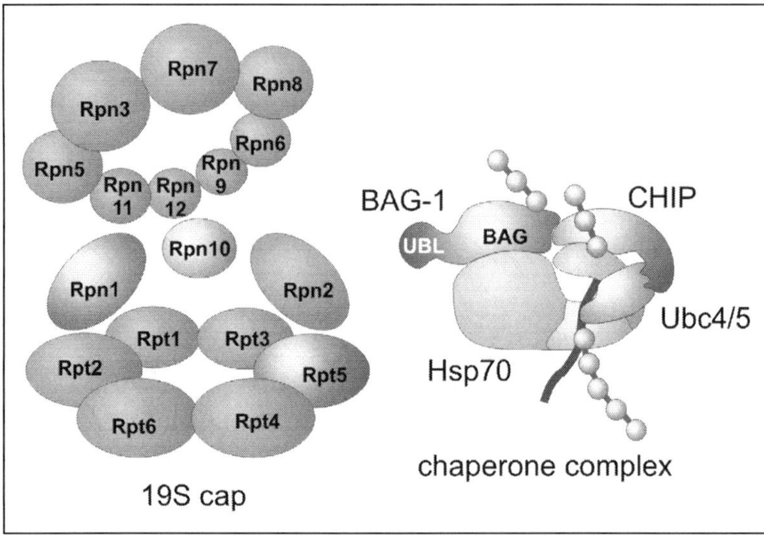

Figure 6. Subunit arrangement of the 19S regulatory particle of the proteasome and schematic presentation of the BAG-1/Hsp70/CHIP complex. BAG-1 is able to cooperate with CHIP in chaperone-assisted degradation because it uses its BAG domain for binding to the ATPase domain of Hsp70 while CHIP simultaneously occupies the carboxyl terminus of the chaperone. BAG-1 and Hsp70 are both ubiquitylated by CHIP in the assembled chaperone/co-chaperone complex. The complex would thus expose multiple proteasomal sorting signals, i.e., the integrated ubiquitin-like domain of BAG-1 (UBL) and polyubiquitin chains attached to BAG-1, Hsp70 and the chaperone client. The UBL domain is recognized by the Rpn1 subunit of the 19S cap. Rpn10 and Rpt5 have been implicated in ubiquitin chain binding. A color version of this figure is available online at www.eurekah.com.

chains are formed with a preference for lysine-27.[40-42] Reduced affinity for such chains could possibly be compensated by the multiple sorting signals exposed by the delivery complex. Furthermore, direct contacts between delivery factors and subunits of the regulatory cap might be essential for coordinating substrate transfer and insertion into the proteasome core.

A Novel Concept for Protein Quality Control

The described examples illuminate a close cooperation between molecular chaperones and the ubiquitin/proteasome system. Chaperones are actively involved at the distinct stages of the degradation process, i.e., substrate selection, sorting, and docking at the proteasome, due to their cooperation with co-chaperones that combine chaperone- and degradation-regulating functions. This challenges previous models for protein quality control, which invoked a competition between chaperones and components of the degradation machinery in substrate selection.[43] Instead a novel concept is emerging, in which molecular chaperones act as central players that would initially bind and stabilize a misfolded protein to direct it either towards folding or degradation depending on the associated co-chaperones (Fig. 7).[4,6] Yet, the functional characterization of the co-chaperones described above marks only the beginning of the elucidation of chaperone-assisted degradation. Many more players remain to be investigated. This is best illustrated by the fact that a deletion of the *chip* gene in mice does not result in a phenotype consistent with impaired protein degradation. The mice are unable to mount a heat shock response due to the involvement of CHIP in the regulation of the heat shock transcription factor,[44] but the degradation of chaperone clients in CHIP-deficient cells proceeds normally.[45] This strongly argues for the existence of additional ubiquitin ligases that are able to cooperate

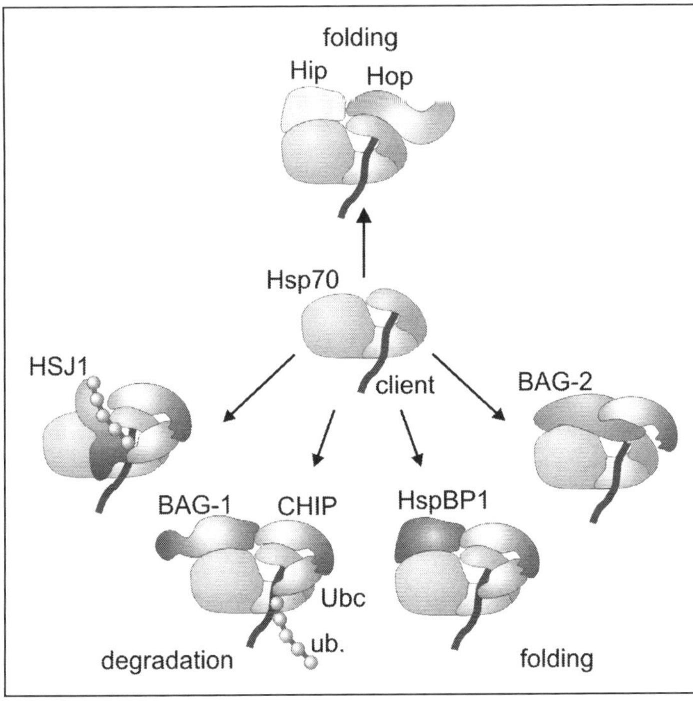

Figure 7. The co-chaperone network that determines protein folding and degradation activities of Hsp70. After initial binding of Hsp70 to a nonnative protein client, diverse co-chaperones can associate with the chaperone to direct the client onto a folding or degradation pathway. CHIP cooperates with HSJ1 and BAG-1 in client degradation. In conjunction with HspBP1 or BAG-2 the ubiquitin ligase activity of CHIP is inhibited. Hip and Hop compete with degradation-inducing co-chaperones in Hsp70 binding and facilitate chaperone-mediated folding. Ub.: ubiquitin chain. A color version of this figure is available online at www.eurekah.com.

with molecular chaperones. A candidate for such a ligase is parkin, the inactivation of which causes early onset Parkinson's disease.[46] Parkin is a RING type ubiquitin ligase that also possesses a UBL domain for proteasome binding. An involvement in chaperone-assisted degradation is indicated by the fact that parkin binds to and cooperates with CHIP and Hsp70 during the ubiquitylation of the disease-associated receptor Pael-R.[47] Although molecular details of this cooperation remain to be explored, parkin may represent an additional pathway for the degradation of chaperone clients and may compensate for a loss of CHIP activity. Functional redundancy may also exist at the level of substrate unloading at the proteasome. Similar to BAG-1, BAG-6/Scythe combines a BAG and an UBL domain and could thus act as a proteasome-associated nucleotide exchange factor of Hsp70 during client delivery.[48] However, experimental evidence for this hypothesis remains to be obtained.

Another chaperone-assisted degradation pathway involves the Cdc48/p97 protein, which belongs to a family of AAA ATPases that form hexameric ring complexes with chaperone-like properties.[15,49] Cdc48/p97 is of central importance for the degradation of misfolded ER proteins that need to be exported into the cytoplasm to reach the proteasome.[50] Export and sorting in the cytoplasm depend on Cdc48/p97 and rely on its chaperone activity. Furthermore, Cdc48/p97 is also required for the degradation of several soluble cytoplasmic proteins.[15] During proteasomal sorting Cdc48/p97 closely cooperates with certain ubiquitin ligases and a broad range of substrate adaptors, many of which contain ubiquitin-binding domains.[14,15,49]

Again, efficient substrate delivery to the proteasome depends on the functional communication between a chaperone system and specialized components of the ubiquitin/proteasome system.

Substrate Selection

Diverse chaperone-assisted degradation pathways probably evolved to mediate the degradation of distinct sets of protein substrates. Indeed, BAG-1 stimulates the CHIP-induced degradation of the glucocorticoid hormone receptor, but does not affect the degradation of CFTR mediated by CHIP.[8,34] Such a substrate selectivity might be explained by differences in the availability of certain co-chaperones at distinct cellular locations. In addition, direct interactions between co-chaperones and the protein substrate are conceivable. In support of this notion, BAG-1, CHIP and HSJ1 were found to bind chaperone clients on their own with various affinities.[16,19] Such direct interactions could contribute to the selection of chaperone clients for proteasomal degradation. It remains to be elucidated what structural features or sequence motifs of the clients are recognized by the degradation-inducing co-chaperones. Identifying such co-chaperone binding sites may help to verify whether the chaperone machinery is able to distinguish between nonnative proteins doomed for degradation and those that need to be refolded. The possibility remains, however, that no such distinction is made during the initial encounter of a nonnative protein with the chaperone machinery. In such a scenario, an irreversibly misfolded protein would cycle on and off the chaperone because of its inability to proceed on its folding pathway and would finally encounter a degradation-inducing chaperone complex, which would direct the protein towards the proteasome.

Regulating the Balance between Chaperone-Assisted Folding and Degradation

Turning a chaperone into a protein degradation factor is potentially dangerous to the cell if one considers that 10-20% of cellular protein transiently associates with chaperones during their de novo synthesis.[51] Apparently, the destructive potential of degradation-inducing co-chaperones such as CHIP should be carefully controlled. Competition with folding-inducing co-chaperones in chaperone binding is an important aspect in this regard. BAG-1, CHIP and HSJ1 all use docking sites on Hsp70, which are also utilized by co-chaperones that stimulate protein folding (Fig. 7).[4] For example, BAG-1 competes with the Hsp70-interacting Hip in binding to the ATPase domain of the chaperone.[35] Hip promotes the folding capacity of Hsp70.[52,53] A similar competition is observed at the carboxyl terminus of Hsp70 between CHIP and the Hsp70/Hsp90-organizing protein Hop that facilitates folding (see Chapter by Daniel et al).[18,54] The intracellular balance of the competing co-chaperones would thus significantly determine folding and degradation activities of the chaperone system (Fig. 7). As Hip and Hop are usually about 5-10 times more abundant than BAG-1 and CHIP,[19,35] folding pathways appear to be favored. The transcriptional regulation of co-chaperone expression is an important subject for future investigations in this regard. Furthermore, the binding affinity of individual co-chaperones for their chaperone partner has to be taken into account here, which may even be subject to alteration by posttranslational modification. This is another aspect that needs to be explored in more detail.

The characterization of the Hsp70 co-chaperone HspBP1 points to an even more intricate mechanism to inhibit the ubiquitin ligase activity of CHIP. Similar to BAG-1, HspBP1 associates with the ATPase domain of Hsp70 to stimulate nucleotide exchange (see Chapter by Brodsky and Bracher).[55,56] Moreover, also HspBP1 is able to form a ternary complex with Hsp70 and CHIP (Fig. 7).[23] In this complex, however, the ubiquitin ligase activity of CHIP is abrogated. HspBP1 most likely blocks ubiquitin attachment sites from the reach of the CHIP ubiquitin ligase. In agreement with a function as a CHIP inhibitor, overexpression of HspBP1 attenuates the CHIP-mediated degradation of CFTR, whereas siRNA-mediated depletion of the co-chaperone accelerates CFTR turnover.[23] The observed inhibitory mechanism could enable CHIP to participate in the regulation of the chaperone cycle of Hsp70 without inducing client degradation. Indeed degradation-independent functions of CHIP were recently described, for example during the regulation of the heat shock transcription factor (HSF).[44] It remains to be seen, however, whether HSF regulation involves a cooperation of CHIP with inhibitors such as HspBP1. Notably, the BAG-1 related co-chaperone BAG-2 also acts as an inhibitor of the CHIP ubiquitin ligase.[57,58] BAG-2 abrogates the interaction between CHIP and its partner E2 enzyme and in this way interferes with CHIP-mediated ubiquitylation (Fig. 7). Furthermore, the co-chaperone BAG-5 was found to inhibit the ubiquitin ligase parkin by binding to Hsp70/parkin complexes.[59] Multiple control mechanisms are apparently in place to define and restrict chaperone-assisted degradation.

Outlook

The analysis of chaperone-assisted degradation highlights the role of the Hsp70 and Hsp90 co-chaperone network in maintaining a delicate equilibrium between the protection of folding intermediates to ensure the adoption of the native state and the efficient clearance of misfolded species that pose a threat to cell viability. Disruption of this homeostatic balance by genetic mutation, stress or aging has catastrophic consequences. Addressing the many questions that remain will therefore not only provide novel insights into cell biology but will also be of large biomedical relevance.

Acknowledgements

We thank Harmut Scheel, Kay Hofmann and Colin Gordon for sharing unpublished results. Work in the authors' lab was supported by grants of the DFG and BMBF.

References

1. Frydman J. Folding of newly translated proteins in vivo: The role of molecular chaperones. Annu Rev Biochem 2001; 70:603-647.
2. Hartl FU, Hayer-Hartl M. Molecular chaperones in the cytosol: From nascent chain to folded protein. Science 2002; 295:1852-1858.
3. Dobson CM. Principles of protein folding, misfolding and aggregation. Semin Cell Dev Biol 2004; 15:3-16.
4. Höhfeld J, Cyr DM, Patterson C. From the cradle to the grave: Molecular chaperones that may choose between folding and degradation. EMBO Rep 2001; 2:885-890.
5. Hartmann-Petersen R, Seeger M, Gordon C. Transferring substrates to the 26S proteasome. Trends Biochem Sci 2003; 28:26-31.
6. Esser C, Alberti S, Höhfeld J. Cooperation of molecular chaperones with the ubiquitin/proteasome system. Biochim Biophys Acta 2004; 1695:171-188.
7. Hershko A, Ciechanover A. The ubiquitin system. Annu Rev Biochem 1998; 67:425-479.
8. Wolf DH, Sommer T, Hilt W. Death gives birth to life: The essential role of the ubiquitin-proteasome system in biology. Biochim Biophys Acta 2004; 1695:1-2.
9. Pickart CM, Eddins MJ. Ubiquitin: Structures, functions, mechanisms. Biochim Biophys Acta 2004; 1695:55-72.
10. Wolf DH, Hilt W. The proteasome: A proteolytic nanomachine of cell regulation and waste disposal. Biochim Biophys Acta 2004; 1695:19-31.
11. Pickart CM. Mechanisms underlying ubiquitination. Annu Rev Biochem 2001; 70:503-533.
12. Koegl M, Hoppe T, Schlenker S et al. A novel ubiquitination factor, E4, is involved in multiubiquitin chain assembly. Cell 1999; 96:635-644.
13. Thrower JS, Hoffman L, Rechsteiner M et al. Recognition of the polyubiquitin proteolytic signal. EMBO J 2000; 19:94-102.
14. Richly H, Rape M, Braun S et al. A series of ubiquitin binding factors connects CDC48/p97 to substrate multiubiquitylation and proteasomal targeting. Cell 2005; 120:73-84.
15. Elsasser S, Finley D. Delivery of ubiquitinated substrates to protein-unfolding machines. Nat Cell Biol 2005; 7:742-749.
16. Westhoff B, Chapple JP, van der Spuy J et al. HSJ1 is a neuronal shuttling factor for the sorting of chaperone clients to the proteasome. Curr Biol 2005; 15:1058-1064.
17. Ballinger CA, Connell P, Wu Y et al. Identification of CHIP, a novel tetratricopeptide repeat-containing protein that interacts with heat shock proteins and negatively regulates chaperone functions. Mol Cell Biol 1999; 19:4535-4545.

18. Connell P, Ballinger CA, Jiang J et al. The co-chaperone CHIP regulates protein triage decisions mediated by heat-shock proteins. Nat Cell Biol 2001; 3:93-96.
19. Demand J, Alberti S, Patterson C et al. Cooperation of a ubiquitin domain protein and an E3 ubiquitin ligase during chaperone/proteasome coupling. Curr Biol 2001; 11:1569-1577.
20. Jiang J, Ballinger CA, Wu Y et al. CHIP is a U-box-dependent E3 ubiquitin ligase: Identification of Hsc70 as a target for ubiquitylation. J Biol Chem 2001; 276:42938-42944.
21. Murata S, Minami Y, Minami M et al. CHIP is a chaperone-dependent E3 ligase that ubiquitylates unfolded protein. EMBO Rep 2001; 2:1133-1138.
22. Meacham GC, Patterson C, Zhang W et al. The Hsc70 co-chaperone CHIP targets immature CFTR for proteasomal degradation. Nat Cell Biol 2001; 3:100-105.
23. Alberti S, Böhse K, Arndt V et al. The co-chaperone HspBP1 inhibits the CHIP ubiquitin ligase and stimulates the maturation of the cystic fibrosis transmembrane conductance regulator. Mol Biol Cell 2004; 15:4003-4010.
24. Hatakeyama S, Matsumoto M, Kamura T et al. U-box protein carboxyl terminus of Hsc70-interacting protein (CHIP) mediates poly-ubiquitylation preferentially on four-repeat Tau and is involved in neurodegeneration of tauopathy. J Neurochem 2004; 91:299-307.
25. Petrucelli L, Dickson D, Kehoe K et al. CHIP and Hsp70 regulate tau ubiquitination, degradation and aggregation. Hum Mol Genet 2004; 13:703-714.
26. Shimura H, Schwartz D, Gygi SP et al. CHIP-Hsc70 complex ubiquitinates phosphorylated tau and enhances cell survival. J Biol Chem 2004; 279:4869-4876.
27. Sahara N, Murayama M, Mizoroki T et al. In vivo evidence of CHIP up-regulation attenuating tau aggregation. J Neurochem 2005; 94:1254-1263.
28. Esser C, Scheffner M, Höhfeld J. The chaperone-associated ubiquitin ligase CHIP is able to target p53 for proteasomal degradation. J Biol Chem 2005; 280:27443-27448.
29. Hwang JR, Zhang C, Patterson C. C-terminus of heat shock protein 70-interacting protein facilitates degradation of apoptosis signal-regulating kinase 1 and inhibits apoptosis signal-regulating kinase 1-dependent apoptosis. Cell Stress Chaperones 2005; 10:147-156.
30. Cheetham ME, Brion JP, Anderton BH. Human homologues of the bacterial heat-shock protein DnaJ are preferentially expressed in neurons. Biochem J 1992; 284:469-476.
31. Mayer MP, Bukau B. Hsp70 chaperones: Cellular functions and molecular mechanism. Cell Mol Life Sci 2005; 62:670-684.
32. Taylor JP, Hardy J, Fischbeck KH. Toxic proteins in neurodegenerative disease. Science 2002; 296:1991-1995.
33. Takayama S, Sato T, Krajewski S et al. Cloning and functional analysis of BAG-1: A novel Bcl-2-binding protein with anti-cell death activity. Cell 1995; 80:279-284.
34. Takayama S, Bimston DN, Matsuzawa S et al. BAG-1 modulates the chaperone activity of Hsp70/Hsc70. EMBO J 1997; 16:4887-4896.
35. Höhfeld J, Jentsch S. GrpE-like regulation of the hsc70 chaperone by the anti-apoptotic protein BAG-1. EMBO J 1997; 16:6209-6216.
36. Takayama S, Krajewski S, Krajewska M et al. Expression and location of Hsp70/Hsc-binding anti-apoptotic protein BAG-1 and its variants in normal tissues and tumor cell lines. Cancer Res 1998; 58:3116-3131.
37. Takayama S, Xie Z, Reed JC. An evolutionarily conserved family of Hsp70/Hsc70 molecular chaperone regulators. J Biol Chem 1999; 274:781-786.
38. Sondermann H, Scheufler C, Schneider C et al. Structure of a Bag/Hsc70 complex: Convergent functional evolution of Hsp70 nucleotide exchange factors. Science 2001; 291:1553-1557.
39. Lüders J, Demand J, Höhfeld J. The ubiquitin-related BAG-1 provides a link between the molecular chaperones Hsc70/Hsp70 and the proteasome. J Biol Chem 2000; 275:4613-4617.
40. Alberti S, Demand J, Esser C et al. Ubiquitylation of BAG-1 suggests a novel regulatory mechanism during the sorting of chaperone substrates to the proteasome. J Biol Chem 2002; 277:45920-45927.
41. Pickart CM, Cohen RE. Proteasomes and their kin: Proteases in the machine age. Nat Rev Mol Cell Biol 2004; 5:177-187.
42. Alberti S, Demand J, Esser C et al. Ubiquitylation of BAG-1 suggests a novel regulatory mechanism during the sorting of chaperone substrates to the proteasome. J Biol Chem 2003; 278:15702-15703.
43. Wickner S, Maurizi MR, Gottesman S. Posttranslational quality control: Folding, refolding, and degrading proteins. Science 1999; 286:1888-1893.
44. Dai Q, Zhang C, Wu Y et al. CHIP activates HSF1 and confers protection against apoptosis and cellular stress. EMBO J 2003; 22:5446-5458.

45. Xu W, Marcu M, Yuan X et al. Chaperone-dependent E3 ubiquitin ligase CHIP mediates a degradative pathway for c-ErbB2/Neu. Proc Natl Acad Sci USA 2002; 99:12847-12852.
46. Tanaka K, Suzuki T, Hattori N et al. Ubiquitin, proteasome and parkin. Biochim Biophys Acta 2004; 1695:235-247
47. Imai Y, Soda M, Hatakeyama S et al. CHIP is associated with Parkin, a gene responsible for familial Parkinson's disease, and enhances its ubiquitin ligase activity. Mol Cell 2002; 10:55-67.
48. Thress K, Song J, Morimoto RI et al. Reversible inhibition of Hsp70 chaperone function by Scythe and Reaper. EMBO J 2001; 20:1033-1041.
49. Dreveny I, Pye VE, Beuron F et al. p97 and close encounters of every kind: A brief review. Biochem Soc Trans 2004; 32:715-720.
50. Meusser B, Hirsch C, Jarosch E et al. ERAD: The long road to destruction. Nat Cell Biol 2005; 7:766-772.
51. Thulasiraman V, Yang CF, Frydman J. In vivo newly translated polypeptides are sequestered in a protected folding environment. EMBO J 1999; 18:85-95.
52. Höhfeld J, Minami Y, Hartl FU. Hip, a novel co-chaperone involved in the eukaryotic Hsc70/Hsp40 reaction cycle. Cell 1995; 83:589-598.
53. Nelson GM, Prapapanich V, Carrigan PE et al. The heat shock protein 70 co-chaperone hip enhances functional maturation of glucocorticoid receptor. Mol Endocrinol 2004; 18:1620-1630.
54. Demand J, Lüders J, Höhfeld J. The carboxy-terminal domain of Hsc70 provides binding sites for a distinct set of chaperone cofactors. Mol Cell Biol 1998; 18:2023-2028.
55. Raynes DA, Guerriero Jr V. Inhibition of Hsp70 ATPase activity and protein renaturation by a novel Hsp70-binding protein. J Biol Chem 1998; 273:32883-32888.
56. Shomura Y, Dragovic Z, Chang HC et al. Regulation of Hsp70 function by HspBP1: Structural analysis reveals an alternate mechanism for Hsp70 nucleotide exchange. Mol Cell 2005; 17:367-379.
57. Dai Q, Qian SB, Li HH et al. Regulation of the cytoplasmic quality control protein degradation pathway by BAG2. J Biol Chem 2005; 280:38673-38681.
58. Arndt V, Daniel C, Nastainczyk W et al. BAG-2 acts as an inhibitor of the chaperone-associated ubiquitin ligase CHIP. Mol Biol Cell 2005; 16:5891-5900.
59. Kalia SK, Lee S, Smith PD et al. BAG5 inhibits parkin and enhances dopaminergic neuron degeneration. Neuron 2004; 44:931-945.

CHAPTER 11

The Role of Hsp70 and Its Co-Chaperones in Protein Misfolding, Aggregation and Disease

Jacqueline van der Spuy, Michael E. Cheetham* and J. Paul Chapple

Abstract

Molecular chaperones and their associated co-chaperones are essential in health and disease as they are key facilitators of protein folding, quality control and function. In particular, the Hsp70 molecular chaperone networks have been associated with neurodegenerative diseases caused by aberrant protein folding. The pathogenesis of these disorders usually includes the formation of deposits of misfolded, aggregated protein. Hsp70 and its co-chaperones have been recognised as potent modulators of inclusion formation and cell survival in cellular and animal models of neurodegenerative disease. In has recently become evident that the Hsp70 chaperone machine functions not only in folding, but also in proteasome mediated degradation of neurodegenerative disease proteins. Thus, there has been a great deal of interest in the potential manipulation of molecular chaperones as a therapeutic approach for many neurodegenerations. Furthermore, mutations in several Hsp70 co-chaperones and putative co-chaperones have been identified as causing inherited neurodegenerative and cardiac disorders, directly linking the Hsp70 chaperone system to human disease.

Introduction

Molecular chaperone networks have been shown to be fundamentally important to many aspects of human health and disease. In a large number of disease studies, changes in chaperone expression profiles have been observed, such that almost no other class of proteins have been linked to such a large array of human disorders. The Hsp70 family of chaperone proteins, and their co-chaperone regulators, have received particular interest in the field of cancer biology, heart disease and neurodegeneration. Hsp70 biology has not only contributed to our understanding of the molecular mechanism of these conditions, but has also led to the identification of biomarkers for disease states and potential targets for therapeutic intervention.

Given their importance in protein folding and quality control, it is perhaps unsurprising that molecular chaperones have been identified as key modulators of human misfolding disease and in particular neurodegenerations.[1-3] The majority of neurodegenerative disorders,

*Corresponding Author: Michael Cheetham; Division of Molecular and Cellular Neuroscience, Institute of Ophthalmology, UCL, 11-43 Bath Street, London, EC1V 9EL, U.K. Email: michael.cheetham@ucl.ac.uk

Networking of Chaperones by Co-Chaperones, edited by Gregory L. Blatch.

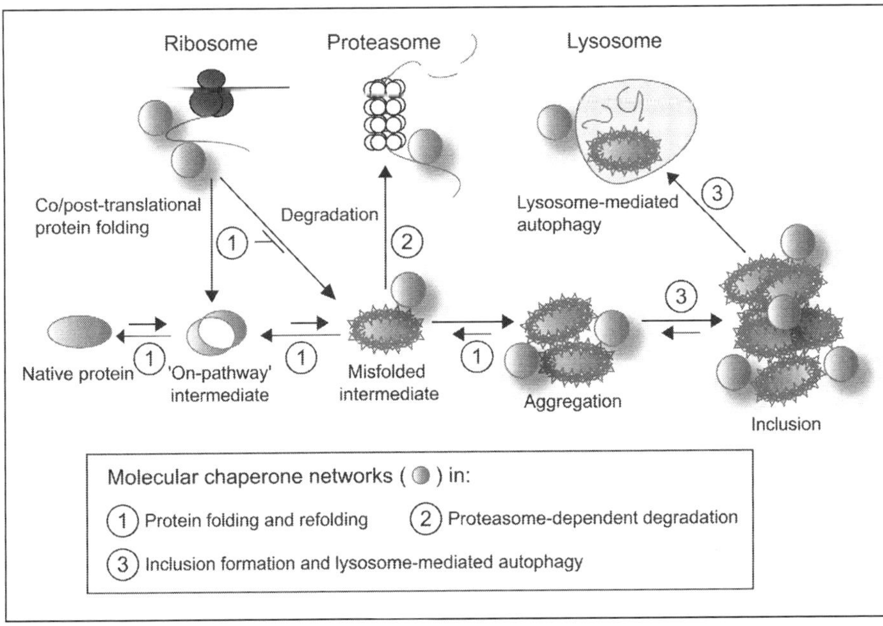

Figure 1. Molecular chaperones in protein misfolding and aggregation. Molecular chaperones and their associated co-chaperones are essential in the cellular defences against protein aggregation. Molecular chaperone networks participate in protein folding and refolding, proteasome-dependent degradation, and inclusion formation and lysosome-mediated autophagy. A color version of this figure is available online at www.eurekah.com.

including Alzheimer's disease, Parkinson's disease, amyotrophic lateral sclerosis (ALS) and polyglutamine (polyQ) expansion diseases, are characterized by conformational changes in proteins that result in misfolding and aggregation.[1,2,4] Some of these aggregates share a propensity to assemble into amyloid fibrils, which are characterised by detergent insolubility, protease resistance, and high β sheet content and cross β sheet structure.[5,6] It has been suggested that during the formation of amyloid fibrils 'off-pathway' assembly may occur resulting in misfolded protein monomers or higher-order aggregates that are not required intermediates in amyloid fibril production.[2] It is unclear why neurons are particularly vulnerable to the accumulation of these off-pathway species, although it has been suggested it may partly be because as post-mitotic cells they can not dilute the toxic proteins during cell division.[2] In neurons and other cells molecular chaperones represent the first line of defence against aberrant protein accumulation (Fig. 1). They are central to the three main cellular defences against protein aggregation;[7] protein folding and refolding;[8,9] proteasome dependent degradation;[10-13] inclusion formation and lysosome-mediated autophagy[14,15] (Fig. 1). Furthermore, folding and proteasomal degradation of proteins are linked through co-chaperones, such as CHIP and HSJ1 (see Chapter by Höhfeld et al),[16] which regulate triage decisions determining whether misfolded proteins are refolded or degraded.

In this chapter we focus on links between the Hsp70 molecular chaperone network and neurodegenerative diseases. Firstly, we consider evidence for the ability of Hsp70 and its co-chaperones to act as suppressors of neurodegeneration, with an emphasis on polyQ misfolding diseases. Secondly, we look at a direct link between the Hsp70 chaperone machine and disease by considering co-chaperones and putative co-chaperones that are mutated in human genetic disorders.

Hsp70 and Its Co-Chaperones in Neurodegenerative Disease

In misfolding disease the ubiquitin proteasome system (UPS) can become compromised and/or overloaded, resulting in ubiquitylated proteins being sequestered in inclusions (Fig. 1).[17] It is uncertain whether such inclusion bodies are pathogenic, incidental or a beneficial coping response. Recent evidence, such as the observation that in a cellular model of Huntington's disease (HD) inclusion body formation reduces the level of mutant huntingtin and the risk of neuronal death,[18] points toward inclusions being part of the cellular defence mechanism;[7] however, it is clear that the presence of inclusions reveals problems of protein processing and could be viewed as surrogate markers of protein misfolding. The mechanisms of cell death in misfolding diseases are not fully resolved; however, certain evidence implies the toxic species could be soluble precursors of the aggregated proteins, rather than the insoluble fibrillar species that are sequestered into inclusions. For example, in a yeast model of HD cell death occurs before insoluble aggregates can be detected.[19]

Chaperones have been shown to colocalise with protein inclusions in cellular and animal models, as well as the lesions observed in human brain tissue. The specific cohort of chaperones associated with inclusions appears to be disease dependent, presumably because although the inclusions have similar biochemical characteristics, the disease protein and cellular context varies. The small heat shock protein (sHSP), the Hsp70 and Hsp40/DnaJ families of molecular chaperones have been most frequently associated with misfolding disease. It has been proposed that the interaction of molecular chaperones and other components of the cellular protein quality control machinery with misfolded proteins may deplete them sufficiently that their normal cellular functions are impaired.[20] Other essential cellular proteins, such as transcription factors, are also recruited to inclusions and this may be detrimental to cell survival.[19,21,22] Of the proteins which are known to be recruited to inclusions molecular chaperones appear to be unique, as they also have the ability to modulate the formation of the inclusions and cell survival. Interestingly, Hsp70 has been demonstrated to be transiently associated with polyQ protein aggregates, exhibiting rapid kinetics of association and dissociation, raising the possibility it may be involved in a pathway rescuing sequestered transcription factors and/or other essential cellular proteins.[23]

The Hsp70 Chaperone Machine

The Hsp70 chaperone machine is a key component of the cellular protein production and quality control machinery. The frequent association of Hsp70 proteins with inclusions of misfolded disease protein suggests this chaperone machine is particularly important in dealing with toxic misfolding disease proteins. Hsp70 proteins bind short regions of peptides with a certain position and pattern of hydrophobic residues in a substrate-binding pocket, assisting in their stabilisation and folding.[8,9] Substrate binding is cyclic with Hsp70 switching from a low substrate affinity, fast substrate exchange state when bound to ATP to a high substrate affinity, slow substrate exchange state upon the hydrolysis of ATP to ADP. Hsp70 undergoes a conformational change resulting in closure of its substrate binding pocket upon ATP hydrolysis,[8,9] dependent on interdomain communication via an allosteric mechanism.[24] This cycle is regulated by Hsp70 co-chaperones and in particular Hsp40/DnaJ proteins, which are characterized by a highly conserved 70-amino acid domain called the J-domain.[25] The J-domain interacts with Hsp70 protein, stimulating ATP hydrolysis and altering substrate binding (see Chapter by Rosser and Cyr).

More than ten Hsp70 genes have been identified in humans, coding for proteins including the cytosolic constitutive heat shock 70 cognate (Hsc70), several stress inducible forms of Hsp70 and the endoplasmic reticulum resident glucose-regulated protein 78 (Grp78 or BiP). Many more Hsp40/DnaJ proteins have been identified. As well as stimulating Hsp70 ATPase activity, Hsp40/DnaJ proteins can bind client proteins independently, directly facilitating targeting to Hsp70.[25] Thus Hsp40/DnaJ proteins may provide a mechanism for recruiting the Hsp70 machine to its many cellular roles. Interestingly, some type II Hsp40/DnaJ proteins, such as HSJ1 and MRJ, are expressed at higher levels in the brain than other tissues suggesting a specificity and/or particular requirements for Hsp70 function in neurons.[26,27]

Hsp70 and Hsp40/DnaJ Proteins as Modulators of PolyQ Protein Aggregation and Toxicity

In 1998 Cummings et al demonstrated that molecular chaperones could be potent modulators of polyQ disease.[28] This report showed that in a cellular model of spinocerebellar ataxia type 1 (SCA-1) overexpression of the Hsp40/DnaJ protein, HDJ-2, caused a significant decrease in the incidence of ataxin-1 inclusions. Subsequently, coexpression of HDJ-2 was demonstrated to reduce inclusion incidence in a model of spinal bulbar muscular atrophy,[29] whilst Hsp40/HDJ-1 was shown to reduce ataxin-3 inclusion formation and toxicity.[30] Interestingly, not all the studies of co-chaperone overexpression have shown beneficial effects on protein aggregation and inclusion formation. For example, it has been reported that in COS-7 cells overexpression of HDJ-2 caused increased inclusion formation in a model of HD.[31] This study, however, represents an exception and there are now multiple reports of Hsp40/DnaJ proteins reducing inclusion incidence and toxicity in cellular models of polyQ diseases. It seems likely that Hsp40/DnaJ proteins which are enriched in neuronal tissues, or have a neuronal specific expression profile, may be particularly relevant in neurodegenerative diseases. In particular, HSJ1a has been shown to effectively reduce the incidence of polyQ protein aggregation, dependent on its interaction with ubiquitin interaction motifs (UIMs),[16] whereas the closely related MRJ, which lacks UIMs, has also been shown to suppress polyQ dependent protein aggregation, caspase activity and cellular toxicity.[27] Surprisingly, few studies have analysed the importance of the co-chaperone or independent chaperone activity of co-chaperones for their protective function. It has been shown, however, that HSJ1a required a functional J-domain to prevent polyQ aggregation.[16] These data suggest that the role of Hsp40/DnaJ proteins in suppressing polyQ toxicity may rely on the regulation of the Hsp70 machine, but this remains to be tested for many other co-chaperones.

In the first in vivo investigation of Hsp70's effect on polyQ disease, the amount of neurodegeneration was reduced but was not effected. This study of a *Drosophila* model of spinocerebellar ataxia type 3 was partially rescued by coexpression of Hsp70.[32] Furthermore, an Hsp70 mutant without ATPase activity had a dominant negative effect making neurodegeneration worse. In the same *Drosophila* model Hsp40/HDJ-1 (but not HDJ-2) was able to suppress degeneration and was also observed to have a synergistic effect with Hsp70, again without altering inclusion formation.[33] When another *Drosophila* model was used to screen for genetic factors modifying degeneration caused by expression of polyQ in the fly eye, two Hsp40/DnaJ proteins were identified, dHDJ-1 and dTPR2, which are potentially homologous to human Hsp40/HDJ-1 and tetratricopeptide repeat protein 2.[34]

The ability of Hsp70 to reduce the severity of polyQ mediated degeneration has also been demonstrated in mouse models. For example, when a SCA-1 transgenic model was crossed with a hemizygous model overexpressing Hsp70 at approximately 10-fold normal levels, behavioural and neuropathological symptoms improved.[35] When animals homozygous and hemizygous for Hsp70 overexpression were compared, results suggested Hsp70 ameliorated polyQ pathologies in a dose dependent manner. Not all mouse models of polyQ disease, however, appear to be equally affected by increasing Hsp70 levels. In a mouse model of HD, overexpression of Hsp70 by 5- to 15-fold only had modest effects on disease progression.[36] This variation may reflect differences in chaperone and co-chaperone expression or the disease models studied.

The Role of the Hsp70 Co-Chaperones CHIP and Bag-1 in Modulating PolyQ Protein Aggregation and Toxicity

Other regulatory components of the Hsp70 chaperone machine have also been recognised as potentially playing important roles in the chaperone response to misfolded disease protein. Of particular interest is the co-chaperone CHIP that negatively regulates Hsp70 chaperone activity and acts as an ubiquitin ligase for Hsp70 client proteins (see Chapter by Höhfeld et al). Overexpression of CHIP was reported to suppress the aggregation and cell death caused by expanded polyQ proteins by increasing the ubiquitination and subsequent degradation of mutant protein[37] and in a recent study HD transgenic mice that were haploinsufficient for CHIP display a markedly accelerated disease phenotype.[38] Interestingly, CHIP has been shown to act as a ubiquitin ligase for the Alzheimer's disease protein tau[39-42] and rescued COS-7 cells from phosphorylated tau induced cell death.[42] The Hsp70 co-chaperone Bag-1 has also been shown to contain an integrated ubiquitin-like domain that enables it to recruit Hsp70 chaperone complexes to the proteasome and to protect cells against polyQ induced cell death.[43] The Hsp70 chaperone machine has a number of other co-chaperone regulators and interacting partners; these proteins represent likely modulators of misfolding disease.

Mechanisms of Hsp70 Mediated Neuroprotection

The potential mechanisms by which the Hsp70 chaperone machine is neuroprotective are manifold and complex. It seems likely that the Hsp70 chaperone machine prevents the conversion of native protein species into toxic intermediates and either facilitates their degradation[12,13,26,44] or, instead, pushes them towards a folding pathway where nontoxic disordered aggregates form. The prominent role played by Hsp70 in the removal of toxic protein species by the UPS means that it helps prevent unwanted interactions between misfolded proteins and important cellular proteins such as transcription factors.[19] Furthermore, Hsp70 has been shown to inhibit the initiation and execution of apoptotic pathways. This is potentially important as the mechanism by which neurons die in neurodegenerative diseases is generally apoptotic.[45] Interestingly, the co-chaperone Bag-1 has also been identified as a potent regulator of apoptosis.[46]

Links between Hsp70 and Other Chaperone Machines

In many of its cellular roles the Hsp70 chaperone machine functions in conjunction with other molecular chaperones systems. For example, the modulation of neurodegeneration by Hsp70 chaperones could be performed in concert with the sHSP family, a diverse group of proteins under 40 kDa in size, that include the α-crystallins and Hsp27. The sHSP share a C-terminal domain of approximately 100 amino acids, which mediates assembly into large oligomeric structures. Upon cellular stress it is believed that these oligomers reorganise into smaller, active complexes which interact with misfolded proteins preventing them from aggregating and maintaining them in a state from which they can potentially be refolded or degraded, by the Hsp70 chaperone machine.[47] There is clear evidence that sHSP can modulate models of misfolding disease. For example, Hsp27 has been shown to prevent cellular polyQ toxicity caused by huntingtin.[48] Furthermore, in yeast it has been demonstrated that Hsp26 alters the nature of polyQ aggregation to facilitate reactivation by the chaperones Hsp104 with the assistance of Hsp70 and Hsp40/DnaJ proteins.[49] It should be noted, however, that as yet no mammalian orthologue of Hsp104 has been identified.

Pharmacological Manipulation of Hsp70 and Other Chaperones

The neuroprotective potential of molecular chaperones may be exploited for the treatment of neurodegenerative diseases. Several drugs have been identified that induce the expression of Hsp70 and other chaperones.[50] These include the hydroxylamine derivative bimoclomol and its analogue, arimoclomol, and the benzoquinone ansamycin antibiotic, geldanymycin, and the related radicol. These compounds potentiate chaperone expression by activating heat shock transcription factor Hsf-1.[50] Hydroxylamine derivatives bind Hsf-1 and prolong its binding to the heat shock response element found in the heat shock gene promoters. In contrast, geldanymycin binds to the ATP site on Hsp90 and block its interaction with Hsf-1 and other clients, thereby stimulating the transcription of heat shock proteins. An exciting recent study demonstrated that treatment with arimoclomol delayed disease progression in ALS mice, resulting in a 22% increase in lifespan.[51] This correlated with a slight increase in Hsp27 levels and a significant increase in both Hsp70 and Hsp90 levels in the spinal cord of the treated ALS mice. As chaperones are fundamentally important in many essential cellular processes it would not be surprising if pharmacological interference in their expression had deleterious effects, although as yet none have been reported with arimoclomol. It may, therefore, be more important to target specific co-chaperones, such as HSJ1 proteins, which have been demonstrated to modulate protein aggregation, but are not ubiquitously expressed.

Mutations in Putative Hsp70 Co-Chaperones Which Cause Inherited Disease

Multiple human disorders have been identified that are associated with mutations in genes encoding chaperones or putative chaperones (Table 1). As yet, no mutations associated with disease have been identified in Hsp70 proteins, possibly because these molecular chaperones are so fundamentally important to cellular survival that mutations would be lethal. However, mutations in several Hsp70 co-chaperones and putative co-chaperones have recently been identified as causing disease. The following is a brief description of some of these proteins.

The BiP Nucleotide Exchange Factor SIL1

In the lumen of the ER the Hsp70 family member BiP plays a crucial role in protein folding, protein translocation and quality control (see Chapters by Brodsky and Bracher, and Dudek et al).[52] Mutations in the BiP co-chaperone SIL1 (or BAP, for BiP associated protein) have been identified as causing the multi-system autosomal recessive disorder Marinesco-Sjögren syndrome (MSS; OMIM 248800).[53,54] This disease affects multiple tissues with key features including: cerebellar ataxia, due to Purkinje and granule cell loss; progressive myopathy with muscle replacement by fat and connective tissue; cataracts from infancy; mental retardation and short stature. The 461 amino acid N-glycosylated SIL1 protein contains ER targeting and retention signals.[55] In the ER SIL1 interacts with the ATPase domain of BiP and induces ADP release and subsequent exchange for ATP, thus regulating the chaperones substrate binding cycle.[55] SIL1 mutations in MSS patients include premature stops, frame shifts and splice site mutations.[53,54] Using homology mapping and replacement based on the co-crystal structure of the cytosolic BiP/SIL1 homologs HspBP1 and Hsp70 it seems likely that the majority of SIL1 mutant proteins will be defective in binding to BiP.[53] Two mutations at the C-terminus of SIL1, which are not predicted to affect the interaction with BiP, are likely to cause mislocalisation as they interfere with the protein's ER retention motif. Interestingly, prior to the identification of mutations in SIL1 in MSS, a spontaneous recessive mouse mutation, woozy (*wz*), was identified as being caused by disruption of the mouse SIL1 gene. The *wz* mouse has an overlapping phenotype with MSS, including adult onset ataxia with loss of cerebellar Purkinje cells.[56] Affected cells have intracellular protein inclusions in the ER and nucleus and upregulation of the unfolded protein response.[56] These data suggest that BiP mediated protein folding is compromised in the ER of affected cells. SIL1 appears to be ubiquitously expressed, so it is unclear why only certain cell types are affected in MSS. Possible mechanisms that could explain this differential cell sensitivity could include: lack of a compensatory factor for mutated SIL1; enhanced sensitivity to an impaired ER chaperone machinery or UPR; specialised client protein requirements.[54] The precise basis remains to be defined, yet the identification of mutations in SIL1 highlight the importance of correct chaperone networking in the ER as well as the cytosol.

Table 1. Genetic disorders associated with mutations in putative chaperones or co-chaperones

Protein	Disease	Main Phenotypic Features	Chaperone System	Function/Putative Function	References
Hsp60	Hereditary spastic paraplegia-13	Progressive spasticity Weakness of lower limbs	Type I chaperonin	Mitochondrial protein folding	87
MKKS/BBS6	McKusick-Kaufman syndrome and Bardet-Biedl syndrome-6	Retinal degeneration Obesity Post-axial poldactyly	Type II chaperonin homology	MKKS has been localized centrosome and cilia to the basal body and is important for normal cilia function and cytokinesis	88,89, 90,91
Tubulin specific chaperone E (TBCE)	Sanjad-Sakati and Kenny-Caffrey	Hypoparathyroidism Mental retardation Facial dysmorphism Growth failure	Post-CCT tubulin specific chaperone pathway	Folding of α-tubulin and its heterodimerization with β-tubulin Regulationof microtubules	92,93
RP2	X-linked retinitis pigmentosa-2	Retinal degeneration	Homology to tubulin specific chaperone C (TBCC)	RP2 has been localised to the plasma membrane and is a partial functional homolog of cofactor C; RP2 interacts with ADP ribosylation factor like 3 protein (Arl3)	94,95, 96,97
SIL1/BAP	Marinesco-Sjögren syndrome	Cerebellar ataxia Cataracts Developmental delay Myopathy	Hsp70 co-chaperone	Regulation of BiP mediated protein folding in the ER	53,54,56
DNAJC19 /Tim14	Dilated Cardiomyopathy with Ataxia (DCMA)	Cardiomyopathy Cerebellar ataxia Growth failure	Hsp70 co-chaperone	Translocation of preproteins across the inner membrane of mitochondria	57

Table continued on next page

Table 1. Continued

Protein	Disease	Main Phenotypic Features	Chaperone System	Function/Putative Function	References
Sacsin	Spastic Ataxia of Charlevoix-Saguenay (ARSACS/SACS)	Motor neuropathy Sensory neuropathy Retinal hypermyelination	Putative DnaJ domain and Hsp90 homology	Possible Hsp70 co-chaperone	66
Aryl hydrocarbon receptor interacting protein-like 1 (AIPL1)	Leber congenital amaurosis	Blindness at birth No detectable ERG Retinal dysfunction and degeneration	TPR protein, putative Hsp70/Hsp90 co-chaperone	Modulation of NUB1 nuclear translocation; interaction with and facilitation of protein farnesylation; post-translational synthesis, biogenesis or assembly of phosphodiesterase (PDE) subunits.	68,69,70, 75,81,82, 84,85
αA-Crystallin	Congenital cataracts	Cataracts	Small heat shock protein	Lens protein biogenesis and protection from damage	98,99
αB-Crystallin	Desmin Related myopathy and Congenital cataracts	Skeletal myopathy Cardiomyopathy Cataracts	Small heat shock protein	Chaperone for the assembly of desmin filaments; lens protein biogenesis and protection from damage	100
Hsp27	Charcot-Marie-Tooth	Weakness Limb muscle atrophy	Small heat shock protein	Altered neurofilament assembly in cells cotransfected with mutant protein, implicating intermediate filament chaperone role	101

The Mitochondrial Hsp40/DnaJ Protein DNAJC19

A splice site mutation in the *DNAJC19* gene has been shown to be associated with a rare autosomal recessive disorder, Dilated Cardiomyopathy with Ataxia (DCMA; OMIM 608977).[57] Features of this condition have been characterized as early onset dilated cardiomyopathy with conductance defects, nonprogressive cerebellar ataxia, testicular dysgenesis, growth failure and 3-methylglutaconic aciduria.[57] DNAJC19 has previously been identified as a component of the mitochondrial proteome[58] and is a putative human orthologue of the yeast Hsp40/DnaJ protein Tim14. Tim14 is essential for cell viability in yeast, as it functions as a component of the Tim23 complex (see Chapter by Bursać and Lithgow), a mitochondrial Hsp70 mediated import motor for the translocation of presequence-containing preproteins across the inner membrane of the mitochondria.[59-61] Interestingly, there are significant similarities between the DCMA phenotype and another disease in which abnormal mitochondria and respiratory chain defects are observed, Barth syndrome (BTHS: OMIM 302060). In BTHS the gene for the Tafazzin protein is mutated.[62] The yeast orthologue of Tafazzin, Taz1, is an outer mitochondrial membrane protein that is exposed to the intermembrane space, which functions as an acyltransferase involved in the remodeling of cardiolipin.[63] The role of cardiolopin is unclear, although it is believed to be required for the proper function of proteins/protein complexes in the inner mitochondrial membrane and has been shown to be critical for the biogenesis of respiratory chain supercomplexes.[63] Thus, it has been suggested that inactivation of Taz1 affects both the assembly and stability of respiratory chain complexes in the inner membrane of mitochondria.[63] It seems likely that the DCMA phenotype reflects a defect in mitochondrial protein import; however, whether there is a direct link between the pathways involved in this disease and BTHS is unresolved. Another mitochondrial Hsp40/DnaJ protein Tid1 (DnajA3) has been identified as having the α-subunit of the mitochondrial DNA polymerase γ (Polga) as an interacting partner.[64] Furthermore, polga has been identified as a client of the yeast homolog of Tid1.[65] Mice deficient in Tid1 have a decreased copy number of mitochondrial DNA and develop dilated cardiomyopathy further illustrating the importance of Hsp40/DnaJ proteins in protein folding in the mitochondria.[64]

The Spastic Ataxia Protein Sacsin

Mutations in the *SACS* gene have been identified as causing the inherited ataxia, autosomal recessive spastic ataxia of Charlevoix-Saguenay (SACS/ARSACS: OMIM 270550).[66] SACS is characterised by early onset of neurodegeneration with absent sensory-nerve conduction, reduced motor-nerve velocity and hypermyelination of retinal-nerve fibres. Although this protein was identified as causing SACS in a Canadian population, recent genetic studies have suggested that this form of spastic ataxia may be more common than originally presumed. The *SACS* gene is predicted to encode sacsin, a large (3829 amino acid) multi-domain protein within a single exon. The *SACS* mRNA appears to be widely expressed in human tissues with enrichment in brain and skeletal muscle. The sacsin protein is a putative co-chaperone of the Hsp40/DnaJ family based upon the presence of a J domain at the C-terminus of its predicted amino acid sequence (~60% identity over 30 residues compared to Hsp40/Hdj1). Although the sacsin J domain is divergent from that of Hsp40 it does contain the highly conserved His-Pro-Asp motif. This tripeptide is essential for the stimulation of Hsp70 ATPase activity by Hsp40/DnaJ proteins. Interestingly, the N-terminal half of sacsin contains two regions (~125 amino acids each) of homology to the N-terminal domain of the chaperone Hsp90 (~27% identity to *S. cerevisiae* Hsp90). Hsp90 and the Hdj2 Hsp40/DnaJ protein have been previously implicated to function together in folding pathways previously, for example in the maturation of the glucocorticoid receptor. It has also been proposed that a region of 110 amino acids at the C-terminus of sacsin represents a novel protein domain, the HEPN domain (higher eukaryote and prokaryote nucleotide binding domain).[67] The HEPN domain is found in bacteria and higher eukaryotes, but is believed to be absent from lower eukaryotes, and its function in sacsin is predicted to be nucleotide binding. Although the role of sacsin in the brain is unknown, it is intriguing to speculate that it may be a chaperone for proteins involved in related ataxias.

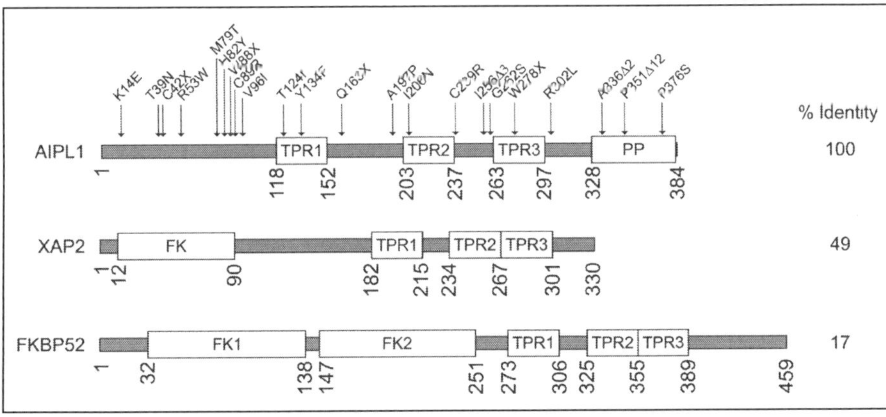

Figure 2. Domain organization of human AIPL1, XAP2 and FKBP52. The domain organizations of human AIPL1 (accession number AAF26708), XAP2 (accession number AAB39923) and FKBP52 (accession number NP_002005) were compared. The known AIPL1 pathogenic mutations are distributed throughout the protein. Human XAP2 shares 49% identity with AIPL1. The level of conservation is highest in the TPR domain, with each XAP2 TPR consensus sharing 50%, 53% and 62% identity respectively with TPR1, TPR2 and TPR3 of AIPL1. A primate-specific proline-rich region (PP) is present at the C-terminus of AIPL1. An FKBP12-like domain (FK) which is unable to bind immunosuppressant drugs and does not have peptidylprolyl isomerase activity is present at the N-terminus of XAP2. This region shares 30% identity with FK1 in FKBP52. The overall degree of conservation between FKBP52 and AIPL1 is 17%, with identity being highest in the TPR domain (27%).

The Aryl Hydrocarbon Receptor Interacting Protein-Like 1 (AIPL1)

Mutations in the *AIPL1* gene cause the autosomal recessive disorder Leber congenital amaurosis (LCA: OMIM 604392), the most severe form of retinal dystrophy characterised by blindness or severe visual impairment at birth.[68] *AIPL1* is expressed in the pineal gland and the specialised, sensory neurons of the retina, the photoreceptors. Within the neuroretina, the expression of AIPL1 protein coincides with the spatiotemporal differentiation and development of the rod and cone photoreceptors, but is restricted to the rod photoreceptors in the adult human retina suggesting a developmental switch in AIPL1 function.[69,70] AIPL1 shares 49% identity with the human X-associated protein 2 (XAP2) or aryl hydrocarbon (Ah) receptor-activated 9 (ARA9), the mouse homologue of which has been designated the Ah receptor-interacting protein (AIP).[71-73] XAP2 and the immunophilins FKBP51/52 participate with the Hsp70-Hsp90 molecular chaperone machinery in the regulation of their respective cognate client proteins, the aryl hydrocarbon receptor (AhR) and the steroid hormone receptors (see Chapter by Cox and Smith). Furthermore, the association of XAP2 with Hsc70 may also facilitate preprotein transfer from Hsc70 to Tom20 and mediate preprotein import in a TPR-dependent manner, suggesting a more general co-chaperone role for XAP2.[74] In addition to the similarity of AIPL1 to XAP2, the conservation of a TPR domain in AIPL1 suggests that AIPL1 may be a member of the family of TPR co-chaperones (Fig. 2).

The client proteins and partner chaperones of AIPL1 still remain to be fully defined, however, a number of AIPL1-interacting proteins have been identified including the NEDD8 ultimate buster protein 1 (NUB1).[75] NUB1 would appear to participate in proteasomal function, via UBL and UBA domains (see Chapter by Höhfeld et al). NUB1 and a larger isoform, NUB1L, associate with two small, ubiquitin-like proteins, NEDD8 and FAT10, and bind the S5a subunit of the 19 S proteasome activator to recruit these ubiquitin-like proteins and their conjugates for proteasomal degradation.[76-80] Though the mechanistic details of NUB1 and NUB1L association with the ubiquitin-like proteins and proteasomal function are becoming clearer, the molecular function of AIPL1 with respect to NUB1 interaction and proteasomal regulation is not understood. AIPL1 is able to modulate the nuclear translocation of NUB1.[81] Furthermore, AIPL1 is able to behave in a chaperone-like manner to suppress the formation of inclusions arising from N- and C-terminal fragments of NUB1.[81] This effect was specific for NUB1 fragments as AIPL1 had no effect on the formation of inclusions by unrelated, aggregation-prone proteins, including the polyQ disease associated Huntingtin-exon 1-Q103. The AIPL1 homologue XAP2 was unable to interact with or modulate NUB1 nucleocytoplasmic distribution and had no effect on the formation of NUB1 fragment inclusions.[81] This suggested that whilst the similarity between AIPL1 and XAP2 correlates with a conserved function in the modulation of nuclear translocation, the specificity for the client protein differs in each case.

In addition to NUB1, it has also been demonstrated that AIPL1 is able to interact with and enhance the post-translational processing of farnesylated proteins, including the Hsp40/DnaJ protein HDJ2.[82] Recently, mouse models of LCA with either the complete or partial inactivation of AIPL1 expression have suggested that AIPL1 may also function as a potential chaperone for cGMP phosphodiesterase (PDE), an essential component of the visual phototransduction cascade.[83-85] In models of AIPL1 LCA, the levels of all three subunits of the cGMP PDE holoenzyme (α, β and γ) were reduced by a post-transcriptional mechanism before the onset of photoreceptor degeneration. AIPL1 may thus be necessary for the biosynthesis, assembly or stabilization of PDE to proteasomal degradation. The PDE-α subunit is farnesylated and mutations that block farnesylation cause degradation of the protein.[86]

Conclusions

The molecular chaperones and their associated co-chaperones are of central importance to protein function from facilitating folding, transport and translocation, through functional maturation to the clearance of misfolded species. Failure of chaperones to fulfil these vital roles may ultimately contribute to a number of devastating human diseases. A number of inherited human disorders have also been associated with mutations in molecular co-chaperones, the modulatory function of which is essential for the normal regulation of the molecular chaperone networks. Therefore, the central importance of the molecular chaperones and their associated co-chaperones in protein misfolding, aggregation and disease makes them a prime target for pharmacological intervention for the treatment of these diseases.

References

1. Barral JM, Broadley SA, Schaffar G et al. Roles of molecular chaperones in protein misfolding diseases. Semin Cell Dev Biol 2004; 15(1):17-29.
2. Muchowski PJ, Wacker JL. Modulation of neurodegeneration by molecular chaperones. Nat Rev Neurosci 2005; 6(1):11-22.
3. Bonini NM. Chaperoning brain degeneration. Proc Natl Acad Sci USA 2002; 99(Suppl 4):16407-16411.
4. Taylor JP, Hardy J, Fischbeck KH. Toxic proteins in neurodegenerative disease. Science 2002; 296(5575):1991-1995.
5. Dobson CM. Protein folding and misfolding. Nature 2003; 426(6968):884-890.
6. Stefani M, Dobson CM. Protein aggregation and aggregate toxicity: New insights into protein folding, misfolding diseases and biological evolution. J Mol Med 2003; 81(11):678-699.
7. Ross CA, Poirier MA. Opinion: What is the role of protein aggregation in neurodegeneration? Nat Rev Mol Cell Biol 2005; 6(11):891-898.
8. Bukau B, Horwich AL. The Hsp70 and Hsp60 chaperone machines. Cell 1998; 92(3):351-366.
9. Hartl FU, Hayer-Hartl M. Molecular chaperones in the cytosol: From nascent chain to folded protein. Science 2002; 295(5561):1852-1858.
10. Alberti S, Demand J, Esser C et al. Ubiquitylation of BAG-1 suggests a novel regulatory mechanism during the sorting of chaperone substrates to the proteasome. J Biol Chem 2002; 277(48):45920-45927.

11. Alberti S, Bohse K, Arndt V et al. The Co-chaperone HspBP1 Inhibits the CHIP ubiquitin ligase and stimulates the maturation of the cystic fibrosis transmembrane conductance regulator. Mol Biol Cell 2004; 15(9):4003-4010.
12. Connell P, Ballinger CA, Jiang J et al. The co-chaperone CHIP regulates protein triage decisions mediated by heat-shock proteins. Nat Cell Biol 2001; 3(1):93-96.
13. Chapple JP, van der SJ, Poopalasundaram S et al. Neuronal DnaJ proteins HSJ1a and HSJ1b: A role in linking the Hsp70 chaperone machine to the ubiquitin-proteasome system? Biochem Soc Trans 2004; 32(Pt 4):640-642.
14. Cuervo AM, Stefanis L, Fredenburg R et al. Impaired degradation of mutant alpha-synuclein by chaperone-mediated autophagy. Science 2004; 305(5688):1292-1295.
15. Cuervo AM. Autophagy: In sickness and in health. Trends Cell Biol 2004; 14(2):70-77.
16. Westhoff B, Chapple JP, van der Spuy J et al. HSJ1 is a neuronal shuttling factor for the sorting of chaperone clients to the proteasome. Curr Biol 2005; 15(11):1058-1064.
17. Bence NF, Sampat RM, Kopito RR. Impairment of the ubiquitin-proteasome system by protein aggregation. Science 2001; 292(5521):1552-1555.
18. Arrasate M, Mitra S, Schweitzer ES et al. Inclusion body formation reduces levels of mutant huntingtin and the risk of neuronal death. Nature 2004; 431(7010):805-810.
19. Schaffar G, Breuer P, Boteva R et al. Cellular toxicity of polyglutamine expansion proteins: Mechanism of transcription factor deactivation. Mol Cell 2004; 15(1):95-105.
20. Sakahira H, Breuer P, Hayer-Hartl MK et al. Molecular chaperones as modulators of polyglutamine protein aggregation and toxicity. Proc Natl Acad Sci USA 2002; 99(Suppl 4):16412-16418.
21. Nucifora Jr FC, Sasaki M, Peters MF et al. Interference by huntingtin and atrophin-1 with cbp-mediated transcription leading to cellular toxicity. Science 2001; 291(5512):2423-2428.
22. McCampbell A, Taylor JP, Taye AA et al. CREB-binding protein sequestration by expanded polyglutamine. Hum Mol Genet 2000; 9(14):2197-2202.
23. Kim S, Nollen EA, Kitagawa K et al. Polyglutamine protein aggregates are dynamic. Nat Cell Biol 2002; 4(10):826-831.
24. Jiang J, Prasad K, Lafer EM et al. Structural basis of interdomain communication in the Hsc70 chaperone. Mol Cell 2005; 20(4):513-524.
25. Cheetham ME, Caplan AJ. Structure, function and evolution of DnaJ: Conservation and adaptation of chaperone function. Cell Stress Chaperones 1998; 3(1):28-36.
26. Chapple JP, Cheetham ME. The chaperone environment at the cytoplasmic face of the endoplasmic reticulum can modulate rhodopsin processing and inclusion formation. J Biol Chem 2003; 278(21):19087-19094.
27. Chuang JZ, Zhou H, Zhu M et al. Characterization of a brain-enriched chaperone, MRJ, that inhibits Huntingtin aggregation and toxicity independently. J Biol Chem 2002; 277(22):19831-19838.
28. Cummings CJ, Mancini MA, Antalffy B et al. Chaperone suppression of aggregation and altered subcellular proteasome localization imply protein misfolding in SCA1. Nat Genet 1998; 19(2):148-154.
29. Stenoien DL, Cummings CJ, Adams HP et al. Polyglutamine-expanded androgen receptors form aggregates that sequester heat shock proteins, proteasome components and SRC-1, and are suppressed by the HDJ-2 chaperone. Hum Mol Genet 1999; 8(5):731-741.
30. Chai Y, Koppenhafer SL, Bonini NM et al. Analysis of the role of heat shock protein (Hsp) molecular chaperones in polyglutamine disease. J Neurosci 1999; 19(23):10338-10347.
31. Wyttenbach A, Carmichael J, Swartz J et al. Effects of heat shock, heat shock protein 40 (HDJ-2), and proteasome inhibition on protein aggregation in cellular models of Huntington's disease. Proc Natl Acad Sci USA 2000; 97(6):2898-2903.
32. Warrick JM, Chan HY, Gray-Board et al. Suppression of polyglutamine-mediated neurodegeneration in Drosophila by the molecular chaperone HSP70. Nat Genet 1999; 23(4):425-428.
33. Chan HY, Warrick JM, Gray-Board et al. Mechanisms of chaperone suppression of polyglutamine disease: Selectivity, synergy and modulation of protein solubility in Drosophila. Hum Mol Genet 2000; 9(19):2811-2820.
34. Kazemi-Esfarjani P, Benzer S. Genetic suppression of polyglutamine toxicity in Drosophila. Science 2000; 287(5459):1837-1840.
35. Cummings CJ, Sun Y, Opal P et al. Over-expression of inducible HSP70 chaperone suppresses neuropathology and improves motor function in SCA1 mice. Hum Mol Genet 2001; 10(14):1511-1518.
36. Hansson O, Nylandsted J, Castilho RF et al. Overexpression of heat shock protein 70 in R6/2 Huntington's disease mice has only modest effects on disease progression. Brain Res 2003; 970(1-2):47-57.
37. Jana NR, Dikshit P, Goswami A et al. Co-chaperone CHIP associates with expanded polyglutamine protein and promotes their degradation by proteasomes. J Biol Chem 2005; 280(12):11635-11640.

38. Miller VM, Nelson RF, Gouvion CM et al. CHIP suppresses polyglutamine aggregation and toxicity in vitro and in vivo. J Neurosci 2005; 25(40):9152-9161.
39. Hatakeyama S, Matsumoto M, Kamura T et al. U box protein carboxyl terminus of Hsc70-interacting protein (CHIP) mediates poly-ubiquitylation preferentially on four-repeat Tau and is involved in neurodegeneration of tauopathy. J Neurochem 2004; 91(2):299-307.
40. Petrucelli L, Dickson D, Kehoe K et al. CHIP and Hsp70 regulate tau ubiquitination, degradation and aggregation. Hum Mol Genet 2004; 13(7):703-714.
41. Sahara N, Murayama M, Mizoroki T et al. In vivo evidence of CHIP up-regulation attenuating tau aggregation. J Neurochem 2005; 94(5):1254-1263.
42. Shimura H, Schwartz D, Gygi SP et al. CHIP-Hsc70 complex ubiquitinates phosphorylated tau and enhances cell survival. J Biol Chem 2004; 279(6):4869-4876.
43. Jana NR, Nukina N. BAG-1 associates with the polyglutamine-expanded huntingtin aggregates. Neurosci Lett 2005; 378(3):171-175.
44. Bailey CK, Andriola IF, Kampinga HH et al. Molecular chaperones enhance the degradation of expanded polyglutamine repeat androgen receptor in a cellular model of spinal and bulbar muscular atrophy. Hum Mol Genet 2002; 11(5):515-523.
45. Garrido C, Schmitt E, Cande C et al. HSP27 and HSP70: Potentially oncogenic apoptosis inhibitors. Cell Cycle 2003; 2(6):579-584.
46. Townsend PA, Stephanou A, Packham G et al. BAG-1: A multi-functional pro-survival molecule. Int J Biochem Cell Biol 2005; 37(2):251-259.
47. Ehrnsperger M, Graber S, Gaestel M et al. Binding of nonnative protein to Hsp25 during heat shock creates a reservoir of folding intermediates for reactivation. EMBO J 1997; 16(2):221-229.
48. Wyttenbach A, Sauvageot O, Carmichael J et al. Heat shock protein 27 prevents cellular polyglutamine toxicity and suppresses the increase of reactive oxygen species caused by huntingtin. Hum Mol Genet 2002; 11(9):1137-1151.
49. Cashikar AG, Duennwald M, Lindquist SL. A chaperone pathway in protein disaggregation. Hsp26 alters the nature of protein aggregates to facilitate reactivation by Hsp104. J Biol Chem 2005; 280(25):23869-23875.
50. Soti C, Nagy E, Giricz Z et al. Heat shock proteins as emerging therapeutic targets. Br J Pharmacol 2005.
51. Kieran D, Kalmar B, Dick JR et al. Treatment with arimoclomol, a coinducer of heat shock proteins, delays disease progression in ALS mice. Nat Med 2004; 10(4):402-405.
52. Kleizen B, Braakman I. Protein folding and quality control in the endoplasmic reticulum. Curr Opin Cell Biol 2004; 16(4):343-349.
53. Senderek J, Krieger M, Stendel C et al. Mutations in SIL1 cause Marinesco-Sjogren syndrome, a cerebellar ataxia with cataract and myopathy. Nat Genet 2005; 37(12):1312-1314.
54. Anttonen AK, Mahjneh I, Hamalainen RH et al. The gene disrupted in Marinesco-Sjogren syndrome encodes SIL1, an HSPA5 co-chaperone. Nat Genet 2005; 37(12):1309-1311.
55. Chung KT, Shen Y, Hendershot LM. BAP, a mammalian BiP-associated protein, is a nucleotide exchange factor that regulates the ATPase activity of BiP. J Biol Chem 2002; 277(49):47557-47563.
56. Zhao L, Longo-Guess C, Harris BS et al. Protein accumulation and neurodegeneration in the woozy mutant mouse is caused by disruption of SIL1, a co-chaperone of BiP. Nat Genet 2005; 37(9):974-979.
57. Davey KM, Parboosingh JS, McLeod DR et al. Mutation of DNAJC19, a human homolog of yeast inner mitochondrial membrane co-chaperones, causes DCMA syndrome, a novel autosomal recessive Barth syndrome-like condition. J Med Genet 2005.
58. Taylor SW, Fahy E, Zhang B et al. Characterization of the human heart mitochondrial proteome. Nat Biotechnol 2003; 21(3):281-286.
59. Mokranjac D, Sichting M, Neupert W et al. Tim14, a novel key component of the import motor of the TIM23 protein translocase of mitochondria. EMBO J 2003; 22(19):4945-4956.
60. Truscott KN, Voos W, Frazier AE et al. A J-protein is an essential subunit of the presequence translocase-associated protein import motor of mitochondria. J Cell Biol 2003; 163(4):707-713.
61. D'Silva PD, Schilke B, Walter W et al. J protein co-chaperone of the mitochondrial inner membrane required for protein import into the mitochondrial matrix. Proc Natl Acad Sci USA 2003; 100(24):13839-13844.
62. Bione S, D'Adamo P, Maestrini E et al. A novel X-linked gene, G4.5. is responsible for Barth syndrome. Nat Genet 1996; 12(4):385-389.
63. Brandner K, Mick DU, Frazier AE et al. Taz1, an outer mitochondrial membrane protein, affects stability and assembly of inner membrane protein complexes: Implications for Barth Syndrome. Mol Biol Cell 2005; 16(11):5202-5214.

64. Hayashi M, Imanaka-Yoshida K, Yoshida T et al. A crucial role of mitochondrial Hsp40 in preventing dilated cardiomyopathy. Nat Med 2005.
65. Duchniewicz M, Germaniuk A, Westermann B et al. Dual role of the mitochondrial chaperone Mdj1p in inheritance of mitochondrial DNA in yeast. Mol Cell Biol 1999; 19(12):8201-8210.
66. Engert JC, Berube P, Mercier J et al. ARSACS, a spastic ataxia common in northeastern Quebec, is caused by mutations in a new gene encoding an 11.5-kb ORF. Nat Genet 2000; 24(2):120-125.
67. Grynberg M, Erlandsen H, Godzik A. HEPN: A common domain in bacterial drug resistance and human neurodegenerative proteins. Trends Biochem Sci 2003; 28(5):224-226.
68. Sohocki MM, Bowne SJ, Sullivan LS et al. Mutations in a new photoreceptor-pineal gene on 17p cause Leber congenital amaurosis. Nat Genet 2000; 24(1):79-83.
69. van der Spuy J, Chapple JP, Clark BJ et al. The Leber congenital amaurosis gene product AIPL1 is localized exclusively in rod photoreceptors of the adult human retina. Hum Mol Genet 2002; 11(7):823-831.
70. van der Spuy J, Kim JH, Yu YS et al. The expression of the Leber congenital amaurosis protein AIPL1 coincides with rod and cone photoreceptor development. Invest Ophthalmol Vis Sci 2003; 44(12):5396-5403.
71. Carver LA, Bradfield CA. Ligand-dependent interaction of the aryl hydrocarbon receptor with a novel immunophilin homolog in vivo. J Biol Chem 1997; 272(17):11452-11456.
72. Kuzhandaivelu N, Cong YS, Inouye C et al. XAP2, a novel hepatitis B virus X-associated protein that inhibits X transactivation. Nucleic Acids Res 1996; 24(23):4741-4750.
73. Ma Q, Whitlock Jr JP. A novel cytoplasmic protein that interacts with the Ah receptor, contains tetratricopeptide repeat motifs, and augments the transcriptional response to 2,3,7,8-tetrachlorodibenzo-p-dioxin. J Biol Chem 1997; 272(14):8878-8884.
74. Yano M, Terada K, Mori M. AIP is a mitochondrial import mediator that binds to both import receptor Tom20 and preproteins. J Cell Biol 2003; 163(1):45-56.
75. Akey DT, Zhu X, Dyer M et al. The inherited blindness associated protein AIPL1 interacts with the cell cycle regulator protein NUB1. Hum Mol Genet 2002; 11(22):2723-2733.
76. Hipp MS, Raasi S, Groettrup M et al. NEDD8 ultimate buster-1L interacts with the ubiquitin-like protein FAT10 and accelerates its degradation. J Biol Chem 2004; 279(16):16503-16510.
77. Kamitani T, Kito K, Fukuda-Kamitani T et al. Targeting of NEDD8 and its conjugates for proteasomal degradation by NUB1. J Biol Chem 2001; 276(49):46655-46660.
78. Kito K, Yeh ET, Kamitani T. NUB1, a NEDD8-interacting protein, is induced by interferon and down-regulates the NEDD8 expression. J Biol Chem 2001; 276(23):20603-20609.
79. Tanaka T, Kawashima H, Yeh ET et al. Regulation of the NEDD8 conjugation system by a splicing variant, NUB1L. J Biol Chem 2003; 278(35):32905-32913.
80. Tanji K, Tanaka T, Kamitani T. Interaction of NUB1 with the proteasome subunit S5a. Biochem Biophys Res Commun 2005; 337(1):116-120.
81. van der Spuy J, Cheetham ME. The Leber congenital amaurosis protein AIPL1 modulates the nuclear translocation of NUB1 and suppresses inclusion formation by NUB1 fragments. J Biol Chem 2004; 279(46):48038-48047.
82. Ramamurthy V, Roberts M, van den AF et al. AIPL1, a protein implicated in Leber's congenital amaurosis, interacts with and aids in processing of farnesylated proteins. Proc Natl Acad Sci USA 2003; 100(22):12630-12635.
83. Dyer MA, Donovan SL, Zhang J et al. Retinal degeneration in Aipl1-deficient mice: A new genetic model of Leber congenital amaurosis. Brain Res Mol Brain Res 2004; 132(2):208-220.
84. Liu X, Bulgakov OV, Wen XH et al. AIPL1, the protein that is defective in Leber congenital amaurosis, is essential for the biosynthesis of retinal rod cGMP phosphodiesterase. Proc Natl Acad Sci USA 2004; 101(38):13903-13908.
85. Ramamurthy V, Niemi GA, Reh TA et al. Leber congenital amaurosis linked to AIPL1: A mouse model reveals destabilization of cGMP phosphodiesterase. Proc Natl Acad Sci USA 2004; 101(38):13897-13902.
86. Qin N, Baehr W. Expression and mutagenesis of mouse rod photoreceptor cGMP phosphodiesterase. J Biol Chem 1994; 269(5):3265-3271.
87. Hansen JJ, Durr A, Cournu-Rebeix I et al. Hereditary spastic paraplegia SPG13 is associated with a mutation in the gene encoding the mitochondrial chaperonin Hsp60. Am J Hum Genet 2002; 70(5):1328-1332.
88. Kim JC, Ou YY, Badano JL et al. MKKS/BBS6, a divergent chaperonin-like protein linked to the obesity disorder Bardet-Biedl syndrome, is a novel centrosomal component required for cytokinesis. J Cell Sci 2005; 118(Pt 5):1007-1020.
89. Ross AJ, May-Simera H, Eichers ER et al. Disruption of Bardet-Biedl syndrome ciliary proteins perturbs planar cell polarity in vertebrates. Nat Genet 2005; 37(10):1135-1140.

90. Katsanis N, Beales PL, Woods MO et al. Mutations in MKKS cause obesity, retinal dystrophy and renal malformations associated with Bardet-Biedl syndrome. Nat Genet 2000; 26(1):67-70.
91. Stone DL, Slavotinek A, Bouffard GG et al. Mutation of a gene encoding a putative chaperonin causes McKusick-Kaufman syndrome. Nat Genet 2000; 25(1):79-82.
92. Martin N, Jaubert J, Gounon P et al. A missense mutation in Tbce causes progressive motor neuronopathy in mice. Nat Genet 2002; 32(3):443-447.
93. Parvari R, Hershkovitz E, Grossman N et al. Mutation of TBCE causes hypoparathyroidism-retardation-dysmorphism and autosomal recessive Kenny-Caffey syndrome. Nat Genet 2002; 32(3):448-452.
94. Schwahn U, Lenzner S, Dong J et al. Positional cloning of the gene for X-linked retinitis pigmentosa 2. Nat Genet 1998; 19(4):327-332.
95. Grayson C, Bartolini F, Chapple JP et al. Localization in the human retina of the X-linked retinitis pigmentosa protein RP2, its homologue cofactor C and the RP2 interacting protein Arl3. Hum Mol Genet 2002; 11(24):3065-3074.
96. Chapple JP, Hardcastle AJ, Grayson C et al. Mutations in the N-terminus of the X-linked retinitis pigmentosa protein RP2 interfere with the normal targeting of the protein to the plasma membrane. Hum Mol Genet 2000; 9(13):1919-1926.
97. Bartolini F, Bhamidipati A, Thomas S et al. Functional overlap between retinitis pigmentosa 2 protein and the tubulin-specific chaperone cofactor C. J Biol Chem 2002; 277(17):14629-14634.
98. Litt M, Kramer P, LaMorticella DM et al. Autosomal dominant congenital cataract associated with a missense mutation in the human alpha crystallin gene CRYAA. Hum Mol Genet 1998; 7(3):471-474.
99. Mackay DS, Andley UP, Shiels A. Cell death triggered by a novel mutation in the alphaA-crystallin gene underlies autosomal dominant cataract linked to chromosome 21q. Eur J Hum Genet 2003; 11(10):784-793.
100. Vicart P, Caron A, Guicheney P et al. A missense mutation in the alphaB-crystallin chaperone gene causes a desmin-related myopathy. Nat Genet 1998; 20(1):92-95.
101. Evgrafov OV, Mersiyanova I, Irobi J et al. Mutant small heat-shock protein 27 causes axonal Charcot-Marie-Tooth disease and distal hereditary motor neuropathy. Nat Genet 2004; 36(6):602-606.

Index

G

H

I

J

K

L

M

N

P